THE PANAVIA TORNADO AT LOW-LEVEL

Cover – Main image
Tornado GR.4 ZA447/DE, of 31 Squadron based at RAF Marham, passing through the Thirlmere valley on 12 September 2005. *(Scott Rathbone)*

Cover – inset top right
Tornado GR.1 ZA609/Z belonging to the RAF's XIII Squadron, seen at Dunmail Raise on 21 October 1997. *(Adrian Walker)*

Cover – inset top left
A Tornado IDS belonging to the Royal Saudi Air Force enters the Tal-y-Llyn valley in the Mach Loop on 18 March 2008. *(Barry Swann)*

Cover – inset centre left
German Air Force Tornado ECR, 4649, seen over Ullswater on 16 September 2022 during a UK deployment. *(Simon Pearson-Cougill)*

Cover – inset bottom left
Tornado F.3 ZE168/FA of XXV(F) Squadron, passes through Dunmail Raise on 13 June 2008 en route to Kemble Air Show. *(Scott Rathbone)*

Rear Cover – Main image
RAF Tornado F.3 ZE791/XY of 111(F) Squadron, passes through Dunmail Raise on 4 August 2003. *(Scott Rathbone)*

Rear Cover – top right
Tornado GR.1B ZA450/FB of the RAF's 12(B) Squadron, seen at Dunmail Raise on 31 August 1995. *(Adrian Walker)*

Rear Cover – top centre
The RAF tradition of applying special tail markings to aircraft during squadron anniversaries is shown here with Tornado GR.4 ZG756, seen wearing special markings for the 14 Squadron 90th Anniversary, as it passes through the Mach Loop on 10 August 2005. *(Scott Rathbone)*

Rear Cover – top left
Another anniversary special was caught at Thirlmere on 26 February 2007, when Tornado F.3 ZG780 passed through carrying markings for the XXV(F) 90th Anniversary. *(Scott Rathbone)*

THE PANAVIA TORNADO AT LOW-LEVEL

THE ULTIMATE PICTORIAL DISPLAY OF THE TORNADO IN ITS ELEMENT

SCOTT RATHBONE

AIR WORLD

THE PANAVIA TORNADO AT LOW-LEVEL
The Ultimate Pictorial Display of the Tornado in its Element

First published in Great Britain in 2024 by
Air World
An imprint of
Pen & Sword Books Ltd
Yorkshire – Philadelphia

Copyright © Scott Rathbone, 2024

ISBN 978 1 39903 304 6

The right of Scott Rathbone to be identified as Author of this work has been asserted by him in accordance with the Copyright, Designs and Patents Act 1988.

A CIP catalogue record for this book is available from the British Library.

All rights reserved. No part of this book may be reproduced or transmitted in any form or by any means, electronic or mechanical including photocopying, recording or by any information storage and retrieval system, without permission from the Publisher in writing.

Typeset by SJmagic DESIGN SERVICES, India.

Printed and bound in India by Replika Press Pvt. Ltd.

Pen & Sword Books Limited incorporates the imprints of Atlas, Archaeology, Aviation, Discovery, Family History, Fiction, History, Maritime, Military, Military Classics, Politics, Select, Transport, True Crime, Air World, Frontline Publishing, Leo Cooper, Remember When, Seaforth Publishing, The Praetorian Press, Wharncliffe Local History, Wharncliffe Transport, Wharncliffe True Crime and White Owl.

For a complete list of Pen & Sword titles please contact

PEN & SWORD BOOKS LIMITED
George House, Units 12 & 13, Beevor Street, Off Pontefract Road, Barnsley, South Yorkshire, S71 1HN, England
E-mail: enquiries@pen-and-sword.co.uk
Website: www.pen-and-sword.co.uk

or

PEN AND SWORD BOOKS
1950 Lawrence Rd, Havertown, PA 19083, USA
E-mail: uspen-and-sword@casematepublishers.com
Website: www.penandswordbooks.com

Contents

Abbreviations	6
Introduction	8
Background	9
Squadrons	11
UKLFS Operations	13
Accidents	22
Operations	24
Retirement	26
The images	29
Acknowledgements	318

Abbreviations

AARA – Air to Air Refuelling Area
AC – Army Co-operation
ADV – Air Defence Variant
AGL – Above Ground Level
ALARM – Air Launched Anti-Radiation Missile
ARTF – Alkaline Removable Temporary Finish
B – Bomber
BAE – British Aerospace
BST – British Summer Time
CAS – Close Air Support
CMU – Combined Maintenance and Upgrade
ETPS – Empire Test Pilots School
F – Fighter
FJWOEU – Fast Jet and Weapons Operational Evaluation Unit
FLIR – Forward Looking Infra Red
GAF – German Air Force
GI – Ground Instruction
GMT – Greenwich Mean Time
GN – German Navy
GR – Ground Attack and Reconaiasance
HARM – High-speed Anti-Radiation Missile
IDS – Interdiction Strike
IS – Islamic State
ISIS – Islamic State of Iraq and Syria

ItAF – Italian Air Force
LFA – Low Flying Area
MLU – Mid Life Update
MRCA – Multi-Role Combat Aircraft
MTA – Military Training Area
OCU – Operational Conversion Unit
OEU – Operational Evaluation Unit
OLF – Operational Low Flying
OP – Operation
OSD – Out of Service Date
OTA – Overland Training Area
QRA – Quick Reaction Alert
R – Reserve
RAE – Royal Aircraft Establishment
RAF – Royal Air Force
RAFG – Royal Air Force Germany
RAPTOR – Reconnaissance Airborne Pod Tornado
RSAF – Royal Saudi Air Force
RTB – Return To Base
RTP – Return To Produce
SAOEU – Strike Attack Operational Evaluation Unit
SEAD – Suppression of Enemy Air Defences
SLIR – Sideways Looking Infra Red
SQN – Squadron

T – Trainer
TIALD – Thermal Imaging Airborne Laser Designator
TSP – Tornado Sustainment Programme
TTA – Tactical Training Area
TTTE – Tri-National Tornado Training Establishment

TWCU – Tornado Weapons Conversion Unit
TWU – Tactical Weapons Unit
UKLFS – United Kingdom Low Flying System
WFU – Withdrawn From Use

Introduction

The final low-level flight of a Royal Air Force (RAF) Tornado to be photographed within the valleys of the United Kingdom Low Flying System (UKLFS) took place on 9 January 2019, when GR.4 ZA612 flew a sortie from RAF Marham, through Cumbria and the Lake District, part of Low Flying Area (LFA)17. The following day saw a pair of GR.4s head to LFA2 situated in the south-west of England, an area generally devoid of any major valleys suitable for photography and therefore rarely attracting low-level aviation photographers. Tornado low-level training was terminated following a directive from Command on Friday 11 January 2019 and the final ever RAF Tornado low-level sorties were carried out as part of the nine-ship retirement flight on 28 February 2019 utilising the LFAs of 5, 6 and 11, around RAF Cranwell and RAF Marham. In the UK, military aircraft are classified as at low-level if operating at 2,000ft and below, so they don't necessarily have to be deep within valleys to be as such, although of course that's where the real excitement lies! The Tornado was an aircraft that had frequented the UKLFS for the best part of forty years and its withdrawal from service with the RAF (the UK being the first of the four operating nations to completely remove the type from its inventory) would see the military low-flying areas

This image provides a good comparison between the Tornado IDS and the Tornado ADV. The lower aircraft is an IDS in the form of the RAF GR.4 variant, while the higher aircraft shows the much sleeker Tornado ADV in the form of an F.3, both aircraft belonging to the RAF's Fast Jet and Weapons Operational Evaluation Unit (FJWOEU), pictured in LFA7 during July 2005. *(Paul Bunch)*

of the UK noticeably quieter. An increasing number of photographers had been gathering in the valleys of the UKLFS over the years in the hope of capturing aircraft at low-level, with the Tornado being high on the list for many. While the retirement of the RAF's Jaguar fleet in 2007 and Harrier fleet at the end of 2010 had already seen traffic reduced significantly, it was the decreasing numbers of Tornados over subsequent years and its eventual withdrawal that led to the largest single loss and effectively brought about the end of an era for the UKLFS.

Background

The Panavia Tornado, developed by the United Kingdom, West Germany and Italy as they looked to replace older types with a new Multi-Role Combat Aircraft (MRCA), was one of the greatest strike fighters of its generation. It was originally designed as a Cold War, all-weather, low-level, multi-role fighter for the three partner nations and would eventually be sold to Saudi Arabia as the only export customer. Several variants emerged: the original strike/reconnaissance version was known as the Interdiction Strike (IDS), which was operated by all air arms, although the RAF referred to theirs as GR, Ground Attack and Reconnaissance, in line with the Jaguar and Harrier designations, the GR.1 being the original variant. Over time the RAF developed their aircraft for additional roles and when broken down, the list of GR.1 variations would consist of the GR.1 (standard strike attack), GR.1A (reconnaissance), GR.1B (maritime strike attack) and GR.1T (twin stick trainer). Demonstrating its versatility, the Tornado went on to be redesigned as an interceptor known as the Tornado Air Defence Variant (ADV), which saw use with the Royal Air Force, the Italian Air Force and the Royal Saudi Air Force. It was originally developed as an interceptor for the RAF as a stopgap that became necessary between the phasing out of the English Electric Lightning and F-4 Phantom, and the introduction of the proposed Eurofighter. Originally designated as the F.2, the jet was soon further developed into the F.3 version after some initial issues with the F.2 design. As with the RAF, the Italian Air Force (ItAF) needed a stop gap between the phasing out of the F-104 Starfighter and the introduction of the Eurofighter and agreed to lease twenty-four ADVs from the RAF to cover the period required. However the ItAF still had a capability gap when the F.3s were returned and a number of F-16s were acquired until the Eurofighter finally entered service.

Back to the RAF GR fleet and the GR.1 Mid-Life Update (MLU) saw 145 airframes upgraded to GR.4 standard between 1996 and 2003 and with the exception of the maritime attack 'B' variant, which was phased out, the designations remained the same with GR.4, GR.4A and GR.4T. It should be noted though that although the 'A' designation remained, the dedicated reconnaissance role of the 'A' variant had been passed on to other types and those aircraft designated as 'A' versions reverted to the standard strike attack role. Of the 145 GR.1s upgraded, 142 were frontline aircraft for the RAF while three remained with BAE Systems at their Warton facility as test and development aircraft.

A total of 990 Tornados were built, 745 IDS variants along with 194 ADVs, of which 165 were for the RAF.

A further fifty-one aircraft were produced to be used as Suppression of Enemy Air Defences (SEAD) aircraft, these would be designated as the Electronic Combat/Reconnaissance (ECR) variant and could usually be seen carrying High-speed Anti Radiation Missiles (HARM). The ECR would be used by the German Air Force with thirty-five airframes and the Italian Air Force with sixteen converted IDS airframes. Of the IDS variant, 210 were destined for the German Air Force and 112 for the German Navy, with 100 for the Italian Air Force and an eventual total of ninety-six for the Royal Saudi Air Force. The RAF received the first of their 228 GR.1s in 1979 and from 1 July 1980, they would send nineteen aircraft to RAF Cottesmore in Rutland, which, from 29 January 1981, would see all initial training by the three partner nations taking place with the Tri-National Tornado Training Establishment (TTTE). This continued for the Royal Air Force, (West) German Air Force, (West) German Navy and Italian Air Force until its disbandment during early 1999. The training areas around the UK provided the perfect combination for the type, with the low-flying areas and ranges in particular making it an ideal location for the multi-force facility. With the closure of the TTTE, each nation took their training in-house and thus ended the regular sight of German and Italian aircraft within the UKLFS, although it would not be the last time that their Tornados would grace the valleys of the UK, certainly not in the case of the German Air Force. With the retirement of the Tornado by the RAF in March 2019, it brought to an end forty years of based Tornado flying in the UK, its spiritual home.

With a wealth of books now written on the Tornado, many taking an in-depth look at the development and operation of the type, I'm going to concentrate on low-level images for this book, with a small insight into low-level operations from a photographer's perspective. I believe this will be the first book purely dedicated to images of the Tornado at low-level and with so many out there I'm hoping to bring some of the very best to you in these pages. For many years between the mid-2000s and the eventual retirement of the type in 2019, the growing number of low-level aviation photographers would gather at the popular locations in England, Scotland and Wales, some even preferring to venture to places off the beaten track for something more exclusive, but all in the hope of capturing images of the mighty Tornado! For the majority of UK-based photographers as well as those visiting from abroad, it was the ultimate machine to see and photograph at low-level. I have met many foreign visitors on the hillsides over the years and the first aircraft most would ask about would be the Tornado; indeed, when I started it was the Tornado GR.4 I was out to capture above anything else. Many experienced photographers were attuned to the sound of an approaching Tornado and the loss of the roar of the RB199 engines echoing through the valleys, coupled with the reduced activity within the UKLFS, led some photographers to call time on the hobby, with many of those who had been pursuing the hobby for several years feeling that the prospect of what was to come without it, left them with little motivation or passion for the hobby, proving just how popular the type had been within the low-level photography community. There is certainly no doubt that the valleys within the UKLFS are a lot less interesting now that the familiar sound of an approaching GR.4 has been consigned to memory.

As mentioned above, the operators of the Tornado came from the three partner nations, these being the Royal Air Force, (West) German Air Force, (West) German Navy and Italian Air Force, with the only additional customer being the Royal Saudi Air Force and each of these air arms feature in this book. Although many quality images have been captured of the RAF Tornado GR.4 at low-level, many have come in the last few years of operations when the jets carried limited, if any, squadron markings. In this book I will bring you images dating back to the 1980s and with the exception of the F.2, all variants are included as well as all frontline RAF squadrons that flew the Tornado during its lengthy career. From the large RAF squadron markings on the original camouflaged jets to the smaller markings on grey jets and from a variety of special schemes and anniversary tails to head-on in your face swept-wing action shots, this book aims to cover all aspects of what it was like to see the Tornado in its element, passing through the valleys of the UKLFS during their rigorous training – to the delight of waiting photographers.

Squadrons

In my previous book, *Military Low Flying F-4 to F-35*, I mentioned how the impending introduction of the Tornado led the UK's then Labour Government to review the UK Low Flying System, which resulted in a complete revamp of the UKLFS in 1979 prior to its introduction, such was the projected significant increase in low-level activity. I also mentioned how this did indeed prove to be the case, more than doubling the sorties flown by the Jaguar and other fast jet strike-aircraft. Of course, with the TTTE carrying out initial training for all operators at the time, there was a large amount of flying being carried out in the UK – despite eight of the RAF's frontline squadrons eventually being based in West Germany as part of RAF Germany (RAFG). The squadrons of II(AC), XV, 16 and 20 were based at RAF Laarbruch, with IX(B), 14, 17(F) and 31 being based at RAF Brüggen. These aircraft would often fly from their bases in West Germany to the UK to carry out low-level work, usually as part of a wider sortie and potentially with a stop-off at one of the UK bases. The UK-based squadrons were initially made up of the Tornado Weapons Conversion Unit (TWCU) (later 45 Reserve (R) Squadron) based at RAF Honington and with 27 and 617 Squadrons based at RAF Marham. XIII Squadron would also form at Honington in 1990. Therefore, despite the eight squadrons based in West Germany, Tornado low-level activity within the UKLFS was still plentiful, with multiple passes being noted in LFA17 in Cumbria on many occasions throughout the 80s and 90s. With the fall of the Berlin Wall triggering the end of the Cold War during late 1989, major changes were afoot for RAFG as plans were put in place for a huge revamp of the RAF and British Army bases in what was now back to being known as Germany. These plans were initially delayed by events in the Middle East and the outbreak of the Gulf War, but over the next decade all remaining RAFG squadrons were either disbanded or relocated to UK bases. As far as the Tornado was concerned, the RAF Laarbruch squadrons were the first to succumb during the early '90s with XV, 16 and 20 Squadrons disbanding. II(AC) Squadron relocated to RAF Marham and was eventually joined by XIII from RAF Honington, both gaining GR.1As in a dedicated reconnaissance role. Meanwhile, the two original RAF Marham squadrons of 27 and 617 would re-equip with the GR.1B in the maritime attack role, although 27 Squadron would disband and reform as 12(B) Squadron once the Buccaneers had been retired. Although 12(B) Squadron originally stood up at RAF Marham, it was only temporarily until such time as 617 Squadron received their GR.1Bs and both would then relocate to RAF Lossiemouth in 1994. Also relocating to Lossiemouth was XV(R) Squadron which had taken over as the shadow squadron of the TWCU, formally 45(R) Squadron. It would eventually take on the responsibility of all RAF Tornado conversion training alongside the weapons training when the TTTE disbanded in 1999. With generally four squadrons stationed at RAF bases, this left room for one more squadron at RAF Lossiemouth and two at RAF Marham and the withdrawal from RAF Brüggen would fill the available space. The unfortunate 17(F) Squadron would disband while IX(B) and 31 Squadrons would move to RAF Marham with 14 Squadron taking the remaining space at Lossiemouth. This now saw all Tornado GRs based in the UK and although there were less aircraft overall, the potential for low-level activity was increased.

In contrast, the ADV squadrons spent their entire lives based in the UK at either RAF Coningsby, RAF Leeming or RAF Leuchars, the only exception being a small force of four aircraft based in the Falklands with 1435 Flt. RAF Coningsby hosted V(AC) and 29 Squadrons along with the Operational Conversion Unit (OCU) of 229 OCU, later 65(R) Squadron, which was itself replaced by 56(R) Squadron, while RAF Leeming was home to XI(F), 23 and XXV(F) Squadrons. RAF Leuchars meanwhile hosted 43(F) and 111(F) Squadrons, these being joined by 56(R) in March 2003, when F.3 operations were ended at RAF Coningsby as the base was prepared to become the initial Typhoon operating base.

In addition to the frontline and training squadrons, all UK variants were operated in a test role throughout their time in service. A number of airframes were flown by British Aerospace, later BAE Systems, from their facility at Warton in Lancashire. These were generally development aircraft for future upgrades to weapons systems and avionics and were very active at the height of Tornado operations. As well as the manufacturer's aircraft a number of other units existed to evaluate the type throughout their service life. These consisted of 32 Joint Trials Unit (JTU), based at NAS China Lake, California, the Royal Aircraft Establishment (RAE), the Empire Test Pilots School (ETPS) and Tornado Operational Evaluation Unit (TOEU), which would become the Strike Attack Operational Evaluation Unit (SAOEU) combined with Harriers and Jaguars, all based at Boscombe Down. The Tornado F.3 Operational Evaluation Unit (F3OEU) also resided at RAF Coningsby and on 1 April 2004 this would merge with the SAOEU and the Air Guided Weapons Operational Evaluation Unit to form the Fast Jet Weapons Operational Evaluation Unit (FJWOEU) which would itself be based at RAF Coningsby. On 1 April 2006 the FJWOEU merged with the Fast Jet Test Squadron (formerly based at Boscombe Down) to become 41(R) Squadron, this after the disbandment of 41(F) Squadron as a Jaguar squadron. With the Tornado GR.4 and F.3 both present on the squadron, this brought about new squadron markings for both variants and as the 2000s progressed they were generally the only squadron flying the Tornado GR.4 that could be guaranteed to be carrying markings, the frontline aircraft having markings removed as they were 'pooled' for operations rather than allocated to specific squadrons.

UKLFS Operations

While any aircraft from any base in the UK could theoretically be seen in any given valley in any Low Flying Area (LFA), there were certain areas and valleys that could generally be relied upon for certain aircraft from certain bases. LFA7 covers the whole of Wales and includes the now famous 'Mach Loop' in southern Snowdonia. This area could always produce anything from anywhere, although your chances of seeing RAF Marham-based Tornados were always greater than RAF Lossiemouth-based Tornados, for instance. This was mainly due to the distance involved and the fact that the RAF Lossiemouth squadrons had vast areas of Scotland and northern England to utilise closer to home, rather than trekking all the way down to Wales. This also applied to F.3s based at RAF Leuchars, as it was always possible to hear RAF Leuchars-based F.3s leave their home base and head high-level to Wales, before dropping into low-level, doing a circuit of the Loop and pulling out of low-level for a high-level transit back home.

While on the subject of F.3s, and given my experiences over the years, it's maybe worth looking a little closer at their behaviour within the numerous valleys contained in LFA7 and LFA17, when there was still a large contingent of F.3s operating from all three of their original operational bases. Normally, low-level flying wouldn't be something required by an aircraft with a role as an interceptor, as primarily the plan would be to take out threats at a distance using long-range missiles. However, it could be forced into becoming a dog-fighter should the need arise, and this is where low-level training was required in order to combat any threats that might defeat forward defences. With a large number of strike aircraft still using the UKLFS on a regular basis, the F.3 force would generally fly Combat Air Patrols (CAPs) over many of the LFAs contained within it, with sorties usually taking place during the morning and afternoon on most days. These CAPs would take place in Overland Training Areas (OTAs), which were strategically placed in the airspace above the LFAs and which fighter aircraft could utilise to engage aircraft within the valleys of the LFAs. This gave the fighters valuable training in engaging attacking strike aircraft while enabling the strike aircraft to train in fighter avoidance using terrain masking. Unlike the LFAs which were numbered, the OTAs were referred to by letters, with eight areas listed as OTA Alpha through to OTA Hotel. The OTAs were often shared with F-15s and Sea Harriers, but utilised large areas of airspace giving multiple formations the chance to spread out, not only for CAPs, but also for Air Combat Manoeuvres (ACM), which were also trained for within the OTAs. There was a reshuffle of the OTA area lettering during the early 2000s, but the areas covered by them were: North Eastern Scotland, South West Scotland, Northumberland/South East Scotland, Cumbria, North Sea, Wales, South West

England and the Highlands of Scotland. Each OTA had its own radio frequency and with a scanner in Cumbria it was possible to hear the action not only in Cumbria, but Wales and Southern Scotland too. Although I didn't get to see it happen too often, the sight of a Tornado F.3 or two bearing down on an unsuspecting mudmover (a term used to describe strike aircraft at low-level) as it weaves its way through the valleys, has to be one of the most exhilarating experiences of my life – especially when said F.3s dived into the valley literally over my head! Witnessing that happen in the M6 Pass in 2002 is something I can still picture and despite having my camera in my hands I was so enthralled by what I was watching that I failed to raise my camera quickly enough! The poor Jaguar T.4 they located wouldn't have stood a chance as the pair of F.3s chased it out of the valley. One other similar F.3 sighting that I can recall came while carrying out my job as a train driver. While heading back south from Carlisle one afternoon, I was passing through an area known as Southwaite, not too far south of Carlisle between the northern Lake District and the Eden Valley. As I passed through, to my left were a pair of Tornado F.3s attacking none other than a Tucano! While the Tucano appeared to be going as fast as it could, the two F.3s were lumbering around with wings fully forward looking like they were almost on the verge of stalling. Although not taking place within a valley, they were all well below 2,000ft; from memory I would have put them probably at around 1,000ft, so well within the realms of low-level.

Onto the LFAs, and in my early years the Mach Loop wasn't the Mecca then that it became in later years. I spent a lot of my time there in 2001–3, but as soon as I'd started venturing into LFA17 from 2002 onwards, it was clear that frontline types were far more common in LFA17 than in the Loop. I caught a small number of F.3s in the Loop in those early years, mainly from RAF Coningsby, but as soon as I started spending more time in Cumbria it became apparent that they spent much more time up there. As far as the valleys suitable for photography were concerned, the M6 Pass, officially known as the Lune Gorge, was a particular favourite for the F.3s from both RAF Leeming and RAF Coningsby. While this could make for some interesting action above the valley, it could also lead to some epic action within it, whether that be watching the patrolling jets chasing unsuspecting aircraft out of the valley, or watching the CAP'ing aircraft themselves make several runs through the valley when things were quiet. On one occasion I witnessed an afternoon pair pass through the M6 Pass three times in half an hour, which helped to bulk my numbers up that day. Once the F.3s had vacated RAF Coningsby then it made a noticeable difference to traffic levels around the M6 Pass and was the start of a demise that would lead to that particular valley losing significant amounts of traffic in the space of just a few years. The RAF Leuchars and RAF Leeming squadrons soldiered on, but while the RAF Leeming squadrons tended to split their time between the M6 Pass and the Lake District valleys, particularly that of Thirlmere, it stood out to me that the RAF Leuchars squadrons always seemed to favour Thirlmere and I can recall only one occasion where I saw RAF Leuchars-based F.3s in the M6 Pass. I was a bigger fan of the Tornado GR myself, but it was always nice to catch an F.3 at low-level and I have many fond memories of doing just that.

Moving onto the GR variant and, despite my best efforts, the GR.4 was the only GR variant I managed to capture at low-level, except for one Saudi IDS out of Warton. The one remaining GR.1, that of GR.1P ZA326 of the ETPS and later QinetiQ, eluded me constantly until it finally retired with seemingly every other low-level photographer at the time having caught it except myself and John Higgins, my regular partner in crime leading up to its retirement. However, I did shoot at least 115 GR.4s at low-level, not a bad haul out of 142 airframes. Had I not taken a sabbatical for a couple of years around 2011/12, then I believe that my tally would have been more like 125–30, maybe more. I did do well with anniversary special schemes though, catching the majority of the 90th anniversary jets of the squadrons that remained by that time. Having shot my first low-level GR.4 during my very first low-level trip, a visit to the Mach Loop in May 2001, that being a single XIII Squadron jet, it took a while before I saw another there.

There are two days in the Loop that stick in my mind from my early days, the first being my first day out with questionable weather. As a novice at the time I had no idea what the limits were for low flying, and on this occasion I arrived to be greeted by a cloud-base touching the tops of the hills. Thinking that it might not be worth my while climbing the hill, I decided to sit in the car and wait to see what happened, my thinking being that the first movement would undoubtedly be a Hawk. At 08.55 I heard a noise and looked up to see a pair of Tornados scream over my car! Realising that my fears were unfounded, I got out of the car to prepare myself for the climb, only to have a Harrier scream over me less than a minute later. It was a big mistake but I had to take it on the chin and use it as a lesson. On another occasion I'd made it until around 14.00 without an aircraft passing through, and assuming all was lost I decided that I might as well pack up and head home. No sooner had I got to a point that was too low down on the hillside for landlocked photos, a Tornado appeared accompanied by a Hawk. This turned out to be a flight from Boscombe Down by an SAOEU Tornado GR.4, and although I did manage to photograph it, it was in the sky and was another lesson in not leaving too early, although I'm not sure that I ever truly learnt that lesson!

Going back to weather conditions, as time passed I was generally out at least once every week, taking a chance in all weathers, and it became apparent that Tornados, particularly GR.4s, could turn up in some extremely poor conditions – so much so that I started to refer to it as 'GR.4 weather'. One memorable occasion at Thirlmere saw a group of us arrive around 08.30 to be greeted by some dire conditions – unfortunately all too common during Lake District mornings. Visibility was terrible and the view down the valley was limited to say the least. As a group we were all so confident of nothing happening that we had all left our cameras firmly packed away in our bags. So imagine our surprise when there was a sudden rumble of jet noise in the valley and out of the murk emerged a flight of three Tornado GR.4s that, amid a barrage of expletives, we were able to identify as RAF Lossiemouth-based jets. To be fair, the light was awful and at that time we were struggling with first generation digital cameras – not the best above 400 ISO, which would have been required. It was a classic example of GR.4 weather though.

My first trip to the M6 Pass in May 2002 produced a pair of XIII Squadron jets, but despite the M6 Pass becoming my valley of choice from that point onwards, with multiple frontline types on most visits, it still didn't produce many Tornado GR.4s, the most numerous type of fighter in the RAF's inventory at that time. My first visit to Thirlmere in early October 2002 produced a number of Harriers and Tornado F.3s, but no GR.4s, although this would soon change. As I started to spend more time at Thirlmere, gradually migrating away from the M6 Pass to spend more time at the Lake District locations and generally using the M6 Pass as a back up should there be any issues further west, it soon became apparent that Thirlmere was the place for GR.4s, with days of four or more occurring frequently. It wasn't until after 2010, when low-level photography had really taken off and when the Mach Loop started to gain its worldwide fame, that the action tended to shift more towards the Loop than the Lakes and then, like the M6 Pass, Thirlmere's activity seemed to drop off a cliff for a few years. This was until late 2018, when effectively the Mach Loop's fame caused issues and saw its use by RAF frontline jets banned by the RAF, a ban that, as I write, has only recently been lifted after almost five years. The one thing I do know is that for me, Thirlmere will always be 'Tornado Alley'. Not only did it draw Tornado crews, but the acoustics of a couple of locations within the valley really enhanced the sound of an approaching GR.4, so much so that you knew what it was before seeing it.

Delving further into the RAF Tornado GR force's low-level habits from a more historical point of view – and prior to me starting low-level photography – I used to spend many hours listening to airband scanners and would often hear the Tornados from RAFG routing into the UK. They would route from Germany with a RAFAIR call sign, the standard call sign for RAF aircraft flying into or out of the UK. Once into UK airspace they would switch to a standard squadron call sign to fly their sortie, before switching back to a RAFAIR call sign for their transit back to Germany. With both low-level and night-flying banned in Germany since the early 1990s, it was a regular occurrence to hear this process being played out, often with the jets working their way up to Yorkshire for a low-level let down at Harrogate Lakes, a series of gravel pits near Knaresborough that are still used to this day as a low-level entry point. From here they would often head west towards LFA17, maybe working their way into Scotland for a lunch/fuel stop at RAF Lossiemouth or RAF Leuchars. Of course this could happen at any time and if they were heading to LFA7 then a routing through the Lichfield Corridor would have been more likely, something most photographers will be used to hearing in more recent years with aircraft routing in from RAF Marham and other east coast bases. While not exclusive to RAFG-based Tornados, it is noticeable from the images that the three-tank fit was regularly used for low-level flights, something that I've only seen a handful of times during my years photographing Tornados at low-level and which coincided with the withdrawal from Germany. Adrian Walker's logbook from LFA17 dating back to the mid-1980s, shows multiple entries of numerous RAFG flights through Thirlmere, as well as flights by UK-based GR.1s. Although rare, flights of six were noted, but it's also worth mentioning that days of traffic without the sight of a GR.1 were also relatively rare. LFA17 was well positioned for all Tornado bases in the UK, being almost centrally located between RAF Marham and RAF Lossiemouth and, with the Spadeadam Range in northern Cumbria, this gave an added reason for flights to route through the Lake District as they routed low-level to the range. I've picked out just a few days between 1988 and 2003 to give an idea of Tornado movements throughout that period. These are just the Tornados on that particular day rather than full logs:

1 November 1988
11.25   Tornado GR.1 x2 17(F) Sqn
11.28   Tornado GR.1 x2 14 Sqn + unmarked
12.35   Tornado GR.1 31 Sqn
12.47   Tornado F.3 229 OCU
13.15   Tornado GR.1 x6 27 Sqn
13.35   Tornado GR.1 x2 617 Sqn + TTTE

5 May 1989
09.45   Tornado IDS TTTE (ItAF)
09.48   Tornado GR.1 TTTE
11.50   Tornado GR.1 x2 TWCU
11.57   Tornado GR.1 14 Sqn
11.59   Tornado GR.1 14 Sqn
13.00   Tornado GR.1 14 Sqn

---

* GAF jets from JBG 31 deployed to RAF Coningsby.

10 May 1989
10.59   Tornado IDS GAF*
11.03   Tornado GR.1 TTTE
11.05   Tornado IDS x2 GAF*
11.35   Tornado IDS TTTE (ItAF)
11.47   Tornado IDS GAF*
15.13   Tornado GR.1 27 Sqn + IDS x2 GAF*

19 July 1994
15.24   Tornado GR.1 x2 TTTE
15.30   Tornado (no further details)
15.40   Tornado GR.1 x2 TTTE
16.01   Tornado IDS x2 TTTE (GAF)
16.10   Tornado GR.1 14 Sqn
16.58   Tornado GR.1 x2 17(F) Sqn
17.13   Tornado GR.1 + IDS TTTE (ItAF)

25 April 1995
09.53   Tornado F.3 V(AC) Sqn
10.43   Tornado GR.1 XIII Sqn
10.51   Tornado GR.1 XIII Sqn
11.12   Tornado GR.1 12(B) Sqn
12.05   Tornado x2 (no further details)
12.20   Tornado GR.1 x2 TTTE
12.43   Tornado F.3 56 Sqn
12.57   Tornado IDS TTTE (ItAF)
13.27   Tornado F.3 x2 V(AC) Sqn
13.29   Tornado GR.1 II(AC) Sqn
14.02   Tornado IDS TTTE (GAF)

1 March 1996
11.02 Tornado F.3 x3 43(F) Sqn
12.18 Tornado F.3 11(F) Sqn
13.08 Tornado IDS TTTE (GAF)
13.59 Tornado IDS TTTE (GAF)
14.34 Tornado IDS TTTE (GAF)
15.00 Tornado GR.1 TTTE
15.17 Tornado GR.1 XIII Sqn

15 April 1997
10.58 Tornado GR.1 XV(R) Sqn
11.11 Tornado GR.1 x2 II(AC) Sqn
11.34 Tornado GR.1 II(AC) Sqn
14.45 Tornado GR.1 XV(R) Sqn
14.52 Tornado GR.1 12(B) Sqn
14.58 Tornado GR.1 TTTE
15.22 Tornado IDS TTTE (GAF)
15.24 Tornado IDS TTTE (GAF)
15.30 Tornado GR.1 XIII Sqn

18 July 2002
10.39 Tornado GR.4 XIII Sqn
11.04 Tornado GR.4 31 Sqn
15.47 Tornado F.3 x2 11(F) Sqn
16.02 Tornado GR.4 x2 12(B) Sqn
16.03 Tornado GR.4 12(B) Sqn
16.08 Tornado GR.4 II(AC) Sqn
16.20 Tornado GR.4 x2 12(B) Sqn
16.21 Tornado GR.4 12(B) Sqn

4 August 2003
10.10 Tornado F.3 111(F) Sqn
10.26 Tornado F.3 111(F) Sqn (same jet as the previous pass)
11.30 Tornado GR.4 14 Sqn
11.54 Tornado GR.4 XV(R) Sqn
11.59 Tornado GR.4 x4 12(B) Sqn + one Operation TELIC scheme
12.17 Tornado GR.4 Operation TELIC scheme
12.18 Tornado GR.4 XV(R) Sqn
12.26 Tornado GR.4 XV(R) Sqn
12.27 Tornado GR.4 XV(R) Sqn
12.41 Tornado GR.4 x3 31 Sqn + two unmarked
15.53 Tornado GR.4 unmarked
16.33 Tornado GR.4 x2 unmarked
16.57 Tornado GR.4 XV(R) Sqn
17.15 Tornado GR.4 12(B) Sqn

This latter listing for 4 August 2003 is a little misleading, as days of any activity of this kind were generally non-existent – let alone with this many Tornados. It did, however, become evident as to why this had happened and the clue was in the sheer number of Scottish-based aircraft, with the RAF Leuchars-based F.3 and GR.4s that predominantly came from RAF Lossiemouth. The day became known among enthusiasts as 'Mad Monday', and while it was an extremely rare and unusual event to have so much traffic through LFA17 at that time, it was not exactly a one-off as it became apparent that the first Monday in August is a Bank Holiday in Scotland, unlike the rest of the UK which takes the last Monday of August as their Bank Holiday. For this reason the Scottish LFAs were unsurprisingly closed to traffic on that day but, in line with the rest of the UK's airbases, the Scottish bases took the last Monday in August as their Bank Holiday and were therefore open during the first Monday. This meant that while all RAF bases were open across the UK, a significant amount of low-flying area was unavailable because the Scottish LFAs were closed, therefore concentrating any traffic into the areas of England and Wales. For the Scottish bases in particular, LFA17 was now a prime candidate to provide the training they required, and on this day they hit the Lake District en masse! What was noticeable though was the number of aircraft that turned right towards Penrith at the end of the Thirlmere valley, the normal routing taking them left towards Bassenthwaite. Further investigation led to the discovery of a Christian convention that takes place for three weeks in Keswick during July and the first few days of August, sometimes overlapping the first Monday of August. The organisers of the event request an avoid of the area for low-flying military aircraft and this was the reason for the unusual right turns on this day, although when the same day was overlapped in subsequent years aircraft tended to avoid the route completely. An example of this would be 2005 when numerous Tornados were sighted to the south of Grasmere routing to the east and west, with a single 617 Squadron aircraft making its way through the Thirlmere valley around midday just as the majority of photographers were leaving to try a different location! The M6 Pass did see some Scottish-based Tornado traffic that day as did the Kirkstone/Ullswater route, but nothing like the eighteen GR.4s and two F.3 passes seen in 2003. The following year (2006) saw a return to something like the 2003 numbers with fourteen GR.4s routing through Thirlmere, although the Keswick convention had finished prior to the Scottish Bank Holiday this year, clearing the way for them to route normally through the Lake District. This was to be the last really good year though, with 2008 being the last of any real note. As the Typhoon started to replace the Tornado at RAF Lossiemouth it became obvious that their low-level requirements were far less than that required for the Tornado and there just wasn't the traffic any more. In fact, I believe that RAF Lossiemouth either closes on the Scottish Bank Holiday now or has a no-fly day, so unfortunately Mad Monday has become just like any other Monday – which is generally nothing to shout about.

One thing I did miss out on was the spectacle of seeing a Tornado GR.4 undertaking Operational Low Flying (OLF) down to a minimum of 100ft in one of the three Tactical Training Areas (TTAs), LFA7(T) in mid-Wales, LFA14(T) in the Scottish Highlands and LFA20(T) in the Scottish Borders. For many years there has been talk and much confusion among those new to low-level photography about the times being published on the MOD Low Flying website, with many claiming that the times relate to LFA7 and the Mach Loop. It is true that there is a timetable posted on there, but this is purely the times that the respective areas mentioned above are available for OLF, it provides no guarantee that aircraft will use it. It certainly has no effect on any other part of the LFA concerned and definitely doesn't relate to the Mach Loop in the case of

LFA7(T). Unfortunately, the Tornado was the last of the RAF fighters to undertake operational low flying, with no requirement for either the Typhoon and F-35 Lightning II. In fact, the last aircraft type to perform OLF in the TTAs was the C-130, which were allowed down to 150ft, but this is yet another type that has now come to the end of its life in RAF service. Time will tell as to whether its replacement, the A400M Atlas, will be permitted to fly down to 150ft at some point. But I digress, and although I never managed to witness them myself, I have included several images from other photographers of Tornado GR.4s in the TTAs. It's worth remembering though that you can't always tell from a photo whether a low-level aircraft is at 500ft, 250ft or 100ft, however it will be mentioned in the caption in the case of an OLF flight. Of the three TTAs, LFA7(T) was seemingly the least used, but apparently didn't offer the same training opportunities as the two Scottish areas, with LFA20(T) being the one most popular with photographers due to a wealth of locations for photography and generally a good chance of traffic (and a far shorter drive for most) whether practising OLF training down to 100ft or general low-level training down to 250ft. One thing of note was that while the OLF areas were active, any non-OLF traffic would be required to fly at 500ft instead of the usual 250ft, in order to deconflict with any possible OLF traffic. While the Tornado crews of 12(B), 14, XV(R) and 617 Squadrons at RAF Lossiemouth generally utilised the area of LFA14(T) due to its proximity to their home base, they could still be seen in LFA20(T) where you'd also be more likely to see Tornado GR.4s of II(AC), IX(B), XIII and 31 Squadrons from RAF Marham. The main route through LFA20(T) is the Selkirk to Moffat valley, a valley approximately thirty miles in length following the A708 road and, one particular favourite location among photographers here was Bowerhope Law, a hill on the south side of the valley on the bend overlooking St Marys Loch, which provided stunning banking-over-water images of aircraft down at 100ft. This was the only location that I personally tried in that valley, and indeed in LFA20(T), but despite visiting around seven times I was still unable to catch anything at 100ft. I did catch a fair amount of traffic on my trips there though, including several Tornado GR.4s, and never had a blank day with no passes. While I must admit that I should have put more effort in, when you're passing through the M6 Pass and close by the Lake District en route, it always seemed to make sense just to head to LFA17 with a better chance of traffic, than to potentially waste time and fuel heading further north with generally a lesser chance of traffic compared to the aforementioned. It's only in more recent times that I've regretted not making more of an effort, especially when you see images from those who did, and in many ways it was just bad luck that I didn't manage to catch an OLF flight in seven attempts.

For those using scanners to listen out for any potential activity, there was slight difference in the best way to listen out for traffic depending on whether you were in LFA7 and the Mach Loop, or LFA17 and the Lake District. I have limited experience of Scotland, but I would put that more in line with LFA17. As far as the Mach Loop goes, you often received your best heads-up of potential traffic by listening to the air traffic control frequencies of either London Military, or Swanwick Military when the latter came into being, bearing in mind that we're talking long before flight radar apps. Due to the location of the Loop in the southern part of Snowdonia in Wales, the majority of traffic would be routing in from the north or east, especially when considering Tornados, both GR.1/4s and F.3s, due to the locations of their bases. In most cases, the aircraft would route high- or medium-level from their bases prior to a low-level let-down, a point to where air traffic control would route an aircraft prior to allowing them to continue en route under Visual Flight Rules (VFR) conditions. With a good radio signal being generally received at most locations around the Loop, you could hear the aircraft transiting the airways corridors, giving their intentions once through the corridors and receiving their low-level let-downs, should low-level be their mission. This was generally the norm for LFA7, but it was also possible for Tornado GR.4s to depart RAF Marham and route low-level all the way from there to LFA7 or LFA17. Tuning in to the squadron air-to-air frequencies could help with this to give an idea of what any aircraft in the vicinity might be up to, and this was the main difference between the Mach Loop and the LFA17 locations. For LFA17, aircraft could route long distances at low-level before hitting many locations within, especially the Lake District, so listening to military air traffic control didn't always help. Anything routing in from the south could usually be heard as they would let-down over Morecambe Bay, but many aircraft would let-down over Yorkshire and route in low-level from the east. This is where having the air-to-air frequencies really helped. My first recollections of hearing air-to-air chatter prior to seeing the said aircraft pass my location was with the frequencies of 12 Squadron and 617 Squadron, both based at RAF Lossiemouth. They were heard on different days around ten minutes before they routed through Thirlmere, neither having been heard on control frequencies. In the case of 12(B) Squadron, it was assumed that hearing them on the air-to-air would mean a pair of aircraft, although the first time I heard them it turned out to be a flight of three! Fortunately, they had a bit of spacing between them and the third was heard approaching before it passed us, we could quite easily have missed it though, with our limited view of approaching aircraft and thinking there was only a pair. A similar thing had happened with the 617 Squadron aircraft, their air-to-air frequency had been active for around ten minutes before they raced through the Thirlmere valley, this time it was just a pair but the second pair of the day after a pair from RAF Marham had passed through a little earlier. On another occasion Thirlmere played host to a pair of XXV(F) Squadron aircraft that passed through, followed one to two minutes later by another pair after some chatter on their air-to-air frequency. Although F.3s were relatively common through Thirlmere, it was rare to see them on CAP there as the civil airways above restricted their use of the area, so they were normally seen as singles or in pairs. This appeared to be a split four-ship playing out a low-level 2v2, although I wasn't able to tell from what I heard on the scanner.

This is just a small selection of memories that I can recall of aircraft on their air-to-air frequencies that gave an indication of something in the area that might just pass by my lens – and did just that. Of course, as mentioned, technology has moved on, with scanners now being used less and less on the hillsides in favour of smart phone aircraft radar apps. These apps show the majority of aircraft, both civil and military, that are airborne around the world. These pinpoint aircraft locations and show what is around, making things easier in some situations and scanners made somewhat redundant. However, use of these apps does rely on a good phone signal, which isn't always possible within the valleys of the UKLFS, and critically aircraft transponders needing to be in the correct mode to show up on these apps.

Accidents

Despite always having a standard NATO Low Level Common frequency to utilise when entering or exiting the LFAs, it was rarely used up until around fifteen years ago, when a couple of incidents forced the implementation of a mandatory call on the low-level frequency to make other aircraft aware of their presence. More recently the UK implemented a trial which introduced a VHF Low Level Common Frequency for use within the UK Low Flying System (UKLFS) to be used alongside the NATO Low Level Common frequency. This enabled GA aircraft to monitor the same frequency as most military aircraft, in particular fast jets. GA aircraft are generally permitted to fly down to 500ft through the LFAs with some exemptions for lower and can therefore encounter military aircraft, so it makes sense that all aircraft concerned should have the ability to be made aware of any traffic they could come across in a low-level environment. There have been several accidents over the years involving low-flying aircraft and, being so numerous, the Tornado is a type that has unfortunately been involved in a number of high-profile incidents. On 17 June 1987 one of the most significant accidents occurred in the Lake District near Keswick, when Tornado GR.1 ZA493, the lead aircraft in a pair of 20 Squadron aircraft flying from RAF Laarbruch in West Germany, collided with Jaguar GR.1 XZ116, the lead aircraft of a pair of Jaguars from 41(F) Squadron at RAF Coltishall. At that time the valleys could be flown in either direction and this was deemed as the cause of the accident, as the angle at which the aircraft approached one another gave little or no view of the other until it was too late; the starboard wing of the Tornado impacted the cockpit of the Jaguar giving the Jaguar pilot no chance of survival, although fortunately both Tornado crewmembers were able to eject, receiving only minor injuries. The accident sparked a review of the UKLFS and the outcome was the cessation of two-way flying in the busiest valleys across the UKLFS, leading to the system we have in place today with one-way flowed valleys, including the three main valleys in LFA17 and the main routes through Wales including the Mach Loop.

While the above accident was one of the most significant, the next unfortunately had a more fatal outcome. On 9 August 1988, Tornado GR.1 ZA329 and Tornado GR.1 ZA593 collided head-on near the village of Blencarn, Cumbria, six miles NNE of Appleby. ZA329 was one of the TTTE aircraft based at RAF Cottesmore and was flying as a singleton on a day visual/night Terrain Following Radar (TFR) low-level sortie. ZA593 was number two in a pair of 617 Squadron aircraft from RAF Marham proceeding on a night-time TFR low-level sortie, entering low-level at Harrogate Lakes. All three aircraft were approaching the Warcop roundabout, a designated area where aircraft approaching from different directions must turn a specified way. The number one of the pair was aware of their number two's position, but at 20.42Z (the Z indicating Zulu time, basically GMT during BST) the number one witnessed a fireball as their wingman collided with the TTTE aircraft at 620ft AGL. As with the previous incident, the collision was attributed to the angle that the two aircraft approached each other and that the TFR hadn't picked up the other aircraft in their beams in order to initiate avoidance. It was also noted that the night low-flying rules allowed both aircraft to be where they were and on their respective headings, but also that neither aircraft made a call on the low-level frequency, something that was required a minute prior to reaching the Warcop roundabout. The surviving crew of the lead 617 Squadron aircraft explained that they believed they were sufficiently far away from the roundabout and, with no turn required, deemed a radio call to be unnecessary and it was concluded that the TTTE crew would have considered the same. All four crew members were tragically killed in this accident and a further review into night-time low-level flying was ordered, with new procedures being implemented the following year.

Another indication of the dangers of low-level flying occurred on 23 June 1993, when Tornado GR.1 ZG754, flying as number two in a pair from RAF Brüggen in Germany, collided with a Bell 206B Jetranger III helicopter, G-BHYW, at Farleton Knott near Kendal, Cumbria. The helicopter was surveying pipelines in the area and had just performed a left-hand orbit to give the observer a good view of work being carried out, levelling out on a northerly heading at approximately 380ft. At 10.49 it was struck by the Tornado, severing the rear boom and tail rotor causing the aircraft to spiral to the ground, killing the pilot and observer. The pair of Tornados had departed RAF Brüggen in Germany to engage in range work on the east-coast ranges prior to a low-level attack transit via LFA17 to RAF Leuchars, where they planned to refuel prior to transiting back to RAF Brüggen. Despite departing late due to a minor technical fault, the number two Tornado had caught up with the lead over the ranges and they were now a pair for the low-level transit. At the time of the collision the Tornado crew were unaware of the impact with the helicopter, instead believing that they'd had a bird strike after hearing a bang and witnessing the significant damage to the nose of their aircraft. The Tornado was able to divert to the airfield at BAE Warton assisted by the number one aircraft and it was only during an inspection of the aircraft at Warton that metal fragments were discovered embedded in the nose, leading to the conclusion that they had been involved in a mid-air collision. The findings indicated that yet again neither crew had spotted the other aircraft, although on this occasion it would appear to have been attributed to distraction rather than the converging angle.

Other accidents have occurred when aircraft have collided with their own wingmen, or flown into the ground or sea while undertaking low-level training, generally during poor visibility. This small selection of incidents all proves just how dangerous low-level flying can be with what is probably the most demanding skill of any type of flying, especially in a high-performance fast jet travelling at 450 knots at 250ft AGL!

Operations

As far as the UK is concerned, the majority of our military aircraft designed for the Cold War never witnessed any operational use, something we should all be thankful for. The Tornado, however, would become one of several types to make their combat debut in the early 1990s – although it wouldn't be the Cold War that saw them cutting their teeth in combat, despite being specifically designed for that eventuality. It was Iraq's invasion of Kuwait in August 1990 and the subsequent 'Gulf War' that followed in early 1991 that prompted the Tornados combat debut, but the low-level role it was designed for and had rigorously trained for in the UKLFS and abroad over the years, still came to the fore. On 17 January 1991 'Operation Desert Storm' was launched, aimed at liberating Kuwait from the invading regime of Saddam Hussein; the United Kingdom was one of thirty-five countries involved in the operation. In fact, the UK committed the largest contingent of any European nation, included in this were Tornado GR.1s from II(AC), IX(B), 14, XV, 16, 17(F), 20, 27, 31 and 617 Squadrons, and Tornado F.3s from V(AC), XI(F), 23, XXV(F), 29 and 43(F) Squadrons. For 'Operation Granby', the UK-named operation within Desert Storm, the RAF's Tornado GR.1s were heavily utilised for low-level strikes in the early days of the campaign – and they paid the price, losing eight aircraft during preparation and operations during the campaign, six of which were during strike missions; this out of a total of fifty-five Allied aircraft losses during the conflict. Painted in a desert pink Alkaline Removable Temporary Finish (ARTF) camouflage, the Tornados were initially tasked with neutralising Iraqi airfields using the JP233 runway denial weapon, which required a low-level delivery and therefore employed the skills gained by the crews during their frequent, and often intense, low-level training. They would eventually switch to a medium-level bombing role using laser-guided bombs targeted by Buccaneers using the Pave Spike laser designator pod, but for many they will always be remembered for their low-level exploits during the fierce early days of the Gulf War. Italian Air Force and Royal Saudi Air Force Tornados also participated in Desert Storm, the Royal Saudi Air Force also employing the JP233 weapon.

For the fleet of RAF Tornados this was just the start of what would become a prolonged and continuous spell of operations in various theatres around the Middle East and even Europe, lasting until their retirement some twenty-eight years later. The end of the conflict saw around a dozen Tornado GR.1s remain in the region, providing cover for Operation Southern Watch and Operation Provide Comfort, while further strikes against Iraq were carried out during Operation Desert Fox in 1998. While cover was still provided in the Iraqi theatre, the Tornado GR.1 was called upon for operations during the Kosovo War in 1999, which also saw German Air Force Tornado ECRs providing SEAD cover for attacking Allied aircraft. This would be the last major action for the RAF's Tornado GR.1s, as by the time of escalations arising yet again in Iraq leading to a second Gulf War and Operation TELIC in 2003, the GR.1 had been through a major upgrade to GR.4 standard and were now predominately delivering precision-guided smart weapons. Because of this there was no return to the desert pink camouflage; instead, an ARTF light grey was worn, although the majority of aircraft still carried nose art as the GR.1s had done during Operation Granby, but with the addition of a TELIC Combat Wing badge on the tail, made up of the squadron insignias of the participating squadrons. A small number of these were caught in the UKLFS both prior to their departure and more so upon their return, albeit by what (at the time) was still a relatively small number of dedicated low-level photographers, until it was washed off over the subsequent months.

The RAF withdrew their Tornado GR.4s from Iraq in June 2009, but the next chapter had already begun with the arrival in early 2009 of a number of GR.4s to Kandahar Airfield, Afghanistan, replacing the Harrier GR.7/9s deployed there since November 2004. The RAF Tornado GR.4s remained in Afghanistan until November 2014 and among the total number of sorties flown were 600 'show of force' passes, a tactic that proved to be successful in deterring Taliban attacks and dispersing personnel by flying low and fast, and was ultimately trained for in the UKLFS. Long after the end of the Cold War, and with weapons and weapon delivery having progressed substantially, there was still a requirement for low-level operations and training, despite constant calls for it to be banned by certain individuals. Between March and September 2011, Tornado GR.4s were involved in strikes on Libya and the enforcement of a no-fly zone as part of Operation Ellamy. By March 2015, Tornados were once again attacking targets within Iraq as operations were stepped up against Islamic State (IS), which were extended to Syria in December 2015 against Islamic State of Iraq and Syria (ISIS). Operation Shader, as it was declared, would be their final operation lasting until 31 January 2019, when the final Tornado sorties were flown, with the eight deployed GR.4s returning from RAF Akrotiri in early February 2019.

Operationally, they will without doubt always be best remembered for their low-level exploits during the first Gulf War, so much so that GR.4T ZG750 was repainted into the iconic desert pink camouflage scheme in 2016 to commemorate the 25th anniversary of the Gulf War and subsequent operations, a fitting tribute and one that was welcomed by the now legion of low-level photographers who clamoured to photograph it in the valleys – many of whom were rewarded for their patience with the crews taking it through the valleys of the UKLFS on a regular basis.

Although far less likely to have required the use of their low-level training, the Tornado ADV also made its combat debut during the Gulf War of early 1991 when eighteen F.3s were deployed to Saudi Arabia. However, despite flying over 2,000 patrol sorties, a lack of capability meant that they were utilised further back from Iraqi airspace, which led to F.3s having no opportunities to actively engage enemy aircraft. A small number of F.3s were maintained in Saudi Arabia to patrol the no-fly zones and then, between 1993 and 1995, Tornado F.3s were used to escort NATO aircraft during Operation Deny Flight over Bosnia. In 1999, Tornado F.3s were once again called upon, this time to perform combat air patrols over Yugoslavia during Operation Allied Force, and they returned to the Middle East in 2003 when 43(F) Squadron and 111(F) Squadron from RAF Leuchars joined XI(F) Squadron and XXV(F) Squadron from RAF Leeming for Operation TELIC against Iraq in the second Gulf War. On this occasion the F.3s were given free reign throughout Iraqi airspace, however; this time, a lack of aerial threats was the reason for limited action. Other than these combat deployments, a small force of four Tornado F.3s were stationed with 1435 Flight at RAF Mount Pleasant in the Falkland Islands, the airframes being rotated between the squadrons back home between 1992, when they replaced F-4 Phantoms and 2009, when they were themselves replaced by Typhoons. The unit provided 24-hour, 365-day air defence cover for the Falkland Islands, South Georgia and South Sandwich Islands.

Retirement
With their replacements on the horizon, changes could be seen coming to the RAF's Tornado fleets. The streamlining of the GR force had taken place through the 1990s into the early 2000s when eleven frontline GR.1 squadrons became seven frontline GR.4 squadrons, and changes to the F.3 force had seen 23 Squadron disband at RAF Leeming in February 1994. Although the forthcoming Typhoon would play a part in replacing both the Tornado F.3 and the GR.4, it was the F.3 that would be first to succumb. With RAF Coningsby set to become the initial home of the Typhoon, it was the Tornado F.3s based there that were set to be hit first, with 29 Squadron disbanding in 1998, standing up again in 2003 as the Typhoon OCU. Next to go was V(AC) Squadron, which disbanded in September 2002, thus leaving the OCU of 56(R) and the F.3 OEU as the only remaining F.3 units at the base until March 2003, when 56(R) Squadron relocated to RAF Leuchars. With the OCU and two frontline squadrons now at RAF Leuchars and two frontline squadrons at RAF Leeming, it was XI(F) Squadron who were next to disband, departing from RAF Leeming in October 2005. They would reform at RAF Coningsby as the second frontline Typhoon squadron and lead multi-role Typhoon squadron. This left just XXV(F) at RAF Leeming and they subsequently disbanded in April 2008, with 56(R) disbanding at RAF Leuchars just a few days later. With 56(R) now disbanded, the training provided by the OCU passed onto 43(F) Squadron, although they themselves would disband in July 2009. As a side note, 43(F) were the squadron I saw the least at low-level of all RAF Tornado squadrons, with just a pair and a single seen on the same day in June 2008. The only other one I saw in 43(F) markings was in actual fact flying out of Warton with BAE Systems.

And so it was left to 111(F) Squadron at RAF Leuchars to have the honour of being the final RAF Tornado F.3 squadron. Having equipped with the type during 1990, they would disband on 22 March 2011 and would be captured at low-level by several photographers during their last low-level flight, although this wasn't quite the end of Tornado F.3 flying in the UK. In 2007, QinetiQ leased four Tornado F.3s from the Ministry of Defence (MOD), based at Boscombe Down for weapons' testing. These would end up being the last F.3s flying in the UK, and having performed their final mission during June 2012, the final three were flown to RAF Leeming for scrapping on 9 July 2012, these also being captured at low-level en route by waiting photographers. Regarding the other two nations flying the F.3, the Italian and Royal Saudi Air Forces, the Italian Air Force returned their Tornado F.3s to the RAF in 2004 having leased thirty-five F-16s from the United States as a further interim until the delivery of the delayed Eurofighter and the Royal Saudi Air Force replaced their Tornado ADVs with Eurofighters in 2011.

As for the RAF Tornado GR force, while continued operations overseas kept the training schedule constant for the crews to rotate, it did mean less aircraft available in the UK for such training and also put a lot of strain on the airframes involved in the operations. The original Out of Service Date (OSD) for the RAF's Tornado GR.4 was 2025, but as with most RAF Fighter types in recent years, this was brought forward significantly to 2019. In hindsight, the remaining fleet were getting a little worse for wear; the continuous operations in the Middle East theatre following the first Gulf War had certainly taken their toll on the airframes, and technical issues seemed to be catching up with them. The majority of the 142 aircraft upgraded to GR.4 standard had spent their entire lives rotating on operations abroad, and the 'T' variants would endure an extensive training schedule back home to keep crews current for the ongoing deployments. Therefore, despite the Mid-Life Update, the aircraft were getting tired, and continuous operations over twenty-eight years were unlikely to have been envisaged – even as the airframes were being put through the upgrade. From the original complement of eleven frontline GR.1 squadrons, the seven frontline squadrons that survived the 1990s, II(AC), IX(B), 12(B), XIII, 14, 31 and 617, re-equipped with the GR.4 between 1998 and 2003, as did XV(R) Squadron who had taken on the weapons training role from the TWCU followed by the conversion training previously undertaken by the TTTE. From 2011 the squadrons would start to be wound down, initially with the disbandment of XIII and 14 Squadrons, both in 2011. The famous 'Dambusters' of 617 Squadron had been resident at RAF Lossiemouth since 1994 and would disband there on 28 March 2014. For years the squadron had been involved in commemorating the anniversary of the Dambusters raids with flypasts at Derwent reservoir in Derbyshire, famous as being one of the regular training grounds of the squadron while practising for Operation Chastise in 1943. These low-level flypasts were nothing new to the area, as part of LFA8 it was an area that could see military low-level flying training being carried out on a regular basis, although it was generally classed as being one of the quieter LFAs. With the anniversary flypasts taking place every five years, 2018 would be the first time that the squadron wasn't available to participate because, despite having reformed at RAF Marham in April 2018 – a month before the 16 May anniversary, they were yet to receive their new F-35B Lightning IIs, the first of which arrived in June 2018. As it happened, it would be November 2019 before an RAF F-35B was seen at low-level in the UKLFS. It was hoped that the Dambusters would return to Derwent with F-35Bs for the 80th anniversary of the raids in 2023, but alas, the commemoration didn't take place and would appear to have now ceased.

The next squadron to disband was 12(B) also at RAF Lossiemouth, on 31 March 2014, just three days after 617 Squadron. The two former GR.1B squadrons in the maritime attack role that had arrived at RAF Lossiemouth together from RAF Marham during the spring of 1994, had now disbanded within days of each other. However, while 617 Squadron had a new venture to look forward to with the F-35B, 12(B) Squadron was to receive something of a renaissance. The next squadron to disband was II(AC) Squadron at RAF Marham, which was scheduled for 31 March 2015, upon their return from Afghanistan participating in Operation Herrick. Tornado operations against ISIS within Operation Herrick were expected to be completed by the end of 2015, but the UK Government determined that this plan was premature and the squadron was required to remain active. It was therefore decided that II(AC) Squadron would still disband, but would do so on 12 January 2015 and be immediately re-formed as 12(B) Squadron, taking on II(AC) Squadron's aircraft and assets, with II(AC) then moving to RAF Lossiemouth to reactivate as a Typhoon squadron. This brought 12(B) back to RAF Marham where it had originally activated as a Tornado GR.1B Squadron after the demise of the Buccaneer (effectively replacing 27 Squadron) in October 1993, prior to moving to RAF Lossiemouth with 617 Squadron as mentioned above.

Following its new lease of life, 12(B) disbanded as a Tornado GR.4 Squadron for the second time in February 2018, meaning the RAF were down to just the final two Tornado squadrons of IX(B) Squadron and 31 Squadron. As the squadrons disbanded, the airframes were either distributed between the remaining squadrons or sent to RAF Leeming for Return To Produce (RTP), where the aircraft were stripped of reusable parts and then scrapped. Despite the reduction in squadrons and numbers of GR.4s between 2011 and their retirement in 2019, low-level activity remained fairly consistent, although for several years between 2012 and 2018 the RAF Marham-based squadrons tended to favour LFA7 – and in particular the Mach Loop – over LFA17 and the Lake District/Cumbria, something of a reversal from the past. This can probably be put down to the Loop gaining its worldwide fame during this period, attracting photographers from around the world and the crews effectively giving the photographers the chance to capture them while undertaking their scheduled training. RAF Lossiemouth crews generally concentrated most of their low-level work in Scotland, particularly the Highlands area of LFA14, although they could sometimes be seen in LFA17 and every now and again would venture as far south as LFA7 and the Mach Loop. Withdrawn airframes being ferried to RAF Leeming often routed low-level on the way, as did aircraft being ferried between RAF Lossiemouth and RAF Marham during the latter days of the Tornados time at RAF Lossiemouth.

And so, as mentioned earlier, the order was put out on Friday 11 January 2019 that Tornado low-level training was to be ceased, and the final low-level training flights by RAF Tornados had now taken place, low-level flying being discontinued some two months prior to the end of flying operations and disbandment. All is not lost for fans of the Tornado though. As I write, the three remaining operators, German Air Force, Italian Air Force and Royal Saudi Air Force, are all still flying the Tornado IDS/ECR, and the German Air Force (and occasionally the Royal Saudi Air Force) have been deploying a number of their Tornados to the UK over the last few years for the Cobra Warrior exercises usually held in March and September. The German Air Force even demonstrated some superb low-level flying through LFA17 during their September 2022 deployment, some of the results featuring towards the end of this book. The heady days of the 1980s/1990s with multiple Tornado sorties a day might be long gone, but for those who were able to witness the Tornado in its element at low-level, the memories and images will last a lifetime. Whether you are one of those that did witness the spectacle, or someone who unfortunately didn't, I hope this book will be a fitting tribute to all variants of the type and will prolong the low-level memories of not only a much-loved aircraft, but also one of the greatest low-level fighters of all time.

The images

As you will see, despite covering all variants of the type, the RAF's Tornado GR.4 understandably dominates the pages of this book, with low-level photography really taking off among aviation enthusiasts during its tenure within the RAF. Of 145 airframes upgraded from GR.1 to GR.4 standard, 142 of them were done so for the RAF, with three remaining with BAE Systems at Warton as development aircraft for future trials and upgrades. With help, I've been able to include images of 135 of the 145 airframes, which considering that several barely flew as GR.4s, is still quite an achievement. As well as those entering early storage, nine aircraft were written off, the first of these being ZA599 on 17 May 2002, when it crashed into the Humber estuary near Brough following an in-flight fire and subsequent control loss, both crew members ejecting without injury. As mentioned earlier, in early 2003 tensions in the Gulf region led to a second Gulf War with a number of Tornado GR.4s being deployed as part of Operation TELIC. On 23 March 2003, Tornado GR.4A ZG710 was shot down while returning to its base at Ali Al Salem in Kuwait, after a successful mission over Iraq, destroying the aircraft and killing both crew members who failed to eject in time. The incident created headlines when it emerged that the aircraft had been shot down by friendly fire after it was mistaken for an incoming anti-radiation missile by a US Army Patriot missile battery.

The table below contains information on the GR.4s, including losses, and also gives the page number(s) where images of each can be found.

Serial	Code		Remarks	Page
ZA365	001	GR4(T)	To RAF Leeming for RTP September 2013	111
ZA367	002	GR4(T)	To RAF Leeming for RTP July 2012	96
ZA369	003	GR4A	To RAF Leeming for RTP August 2017	210
ZA370	004	GR4A	To RAF Leeming for RTP December 2017	213, 274
ZA371	005	GR4A	To RAF Leeming for RTP January 2014	135
ZA372	006	GR4A	To RAF Leeming for RTP December 2016	201
ZA373	007	GR4A	To RAF Leeming for RTP June 2015	99
ZA393	008	GR4	To RAF Leeming for RTP July 2015	174
ZA395	009	GR4A	To RAF Leeming for RTP April 2014	208, 230, 264
ZA398	010	GR4A	Cornwall Aviation Heritage Centre via Manston Fire School	239, 240
ZA400	011	GR4A	To RAF Leeming for RTP May 2016	117
ZA401	012	GR4A	To RAF Leeming for RTP January 2014	104, 105, 163, 259
ZA402		GR4A	BAE Systems development aircraft. To RAF Leeming for RTP July 2013.	197, 231
ZA404	013	GR4A	To RAF Leeming for RTP October 2013	233, 252
ZA405	014	GR4A	To RAF Leeming for RTP February 2016	279
ZA406	015	GR4	To RAF Leeming for RTP May 2017	99, 272
ZA410	016	GR4(T)	To RAF Leeming for RTP November 2013	94
ZA412	017	GR4(T)	To RAF Leeming for RTP July 2015	258
ZA446	018	GR4	Written off on 23 September 2009 after diverting to RAF Leuchars with an engine fire which spread to a major fire. With repairs not authorised, it was moved to RAF Shawbury for storage.	95
ZA447	019	GR4	WFU 2017, to RAF Cosford for GI March 2018	120, 200
ZA449	020	GR4	WFU February 2019	125, 146
ZA452	021	GR4	Preserved Midlands Air Museum, Coventry	232
ZA453	022	GR4	To RAF Leeming for RTP June 2017	145

Serial	Code		Remarks	Page
ZA456	023	GR4	To RAF Leeming for RTP September 2015	277
ZA458	024	GR4	To RAF Leeming for RTP April 2017	168
ZA459	025	GR4	WFU January 2018	175, 293
ZA461	026	GR4	To RAF Leeming for RTP September 2015	273
ZA462	027	GR4	To RAF Leeming for RTP September 2017	107, 289
ZA463	028	GR4	WFU March 2019	227, 260
ZA469	029	GR4	Preserved Imperial War Museum Duxford	165, 191
ZA470	030	GR4	Spent most of its life as a GR.4 in storage. To RAF Leeming for RTP January 2013	No image
ZA472	031	GR4	WFU March 2018	261
ZA473	032	GR4	WFU May 2018	174
ZA491		GR4	Written off on 22 July 2004, prior to numbered codes, after crashing into the North Sea during a training flight. Both crewmembers ejected safely.	86
ZA492	033	GR4	To RAF Leeming for RTP May 2015	253, 256
ZA541	034	GR4(T)	To RAF Leeming for RTP June 2016	95
ZA542	035	GR4	WFU March 2019	83
ZA543	036	GR4	To RAF Leeming for RTP December 2018	97, 143, 173, 295
ZA544	037	GR4(T)	To RAF Leeming for RTP July 2012	87
ZA546	038	GR4	WFU July 2018	101
ZA547	039	GR4	To RAF Leeming for RTP January 2014	246, 247
ZA548	040	GR4(T)	To RAF Leeming for RTP June 2017	280
ZA549	041	GR4(T)	To RAF Leeming for RTP November 2011	82, 164
ZA550	042	GR4	To RAF Leeming for RTP October 2017	128, 149, 257
ZA551	043	GR4(T)	To RAF Leeming for RTP October 2015	142
ZA552	044	GR4(T)	To RAF Leeming for RTP May 2012	167
ZA553	045	GR4	WFU March 2019	287, 296
ZA554	046	GR4	WFU January 2019	283, 300
ZA556	047	GR4	WFU November 2018	114
ZA557	048	GR4	To RAF Leeming for RTP March 2016	204

Serial	Code		Remarks	Page
ZA559	049	GR4	To RAF Leeming for RTP August 2017	288
ZA560	050	GR4	WFU February 2018, to RAF Wittering for GI	286
ZA562	051	GR4(T)	To RAF Leeming for RTP December 2015	116
ZA563	052	GR4	Spent most of its life as a GR.4 in storage. To RAF Leeming for RTP September 2012	No image
ZA564	053	GR4	To RAF Leeming for RTP April 2014	115
ZA585	054	GR4	WFU August 2018, to RAF Cosford for GI October 2018	93
ZA587	055	GR4	WFU March 2019	95
ZA588	056	GR4	WFU August 2018	122
ZA589	057	GR4	To RAF Leeming for RTP September 2015	278
ZA591	058	GR4	RTP RAF Marham June 2018	170
ZA592	059	GR4	To RAF Leeming for RTP February 2017	84
ZA594	060	GR4(T)	To RAF Leeming for RTP April 2017	156
ZA595	061	GR4(T)	To RAF Leeming for RTP April 2014	101
ZA596	062	GR4	Written off on 20 July 2009 while departing Kandahar, Afghanistan, after a fuel leak ignited on take-off and forced the aircraft to overrun the runway. Both crew members ejected safely.	147
ZA597	063	GR4	WFU February 2019	99
ZA598	064	GR4(T)	To RAF Leeming for RTP July 2015	127
ZA599		GR4(T)	Written off on 17 May 2002, prior to numbered codes, after crashing into the Humber estuary with mechanical failure. Both crew members ejected safely.	No image
ZA600	065	GR4	To RAF Leeming for RTP April 2015	236
ZA601	066	GR4	WFU March 2019	118, 241
ZA602	067	GR4(T)	To RAF Leeming for RTP November 2016	190
ZA604	068	GR4(T)	To RAF Leeming for RTP January 2015	123
ZA606	069	GR4	To RAF Leeming for RTP November 2014	268, 269
ZA607	070	GR4	WFU January 2019	297
ZA608	071	GR4	Spent most of its life as a GR.4 in storage. To RAF Leeming for RTP October 2012	No image

Serial	Code		Remarks	Page
ZA609	072	GR4	To RAF Leeming for RTP February 2016	112
ZA611	073	GR4	WFU February 2018	179, 182
ZA612	074	GR4(T)	WFU February 2019	171, 302
ZA613	075	GR4	WFU March 2019	100
ZA614	076	GR4	WFU March 2019	267, 270
ZD707	077	GR4	To RAF Leeming for RTP April 2015	207
ZD708		GR4	BAE Systems development aircraft.	No image
ZD709	078	GR4	To RAF Leeming for RTP February 2017	139
ZD711	079	GR4(T)	WFU 2017, to RAF Honington for GI 2018	94, 222
ZD712	080	GR4(T)	To RAF Leeming for RTP June 2014	262
ZD713	081	GR4(T)	To RAF Leeming for RTP February 2018	282
ZD714	082	GR4	To RAF Leeming for RTP December 2013	152
ZD715	083	GR4	WFU June 2016, to RAF Cosford for GI	215, 238
ZD716	084	GR4	WFU March 2019	131, 301
ZD719	085	GR4	To RAF Leeming for RTP July 2014	101
ZD720	086	GR4	To RAF Leeming for RTP February 2014	243
ZD739	087	GR4	To RAF Leeming for RTP July 2017	109
ZD740	088	GR4	To RAF Leeming for RTP October 2015	144
ZD741	089	GR4(T)	To RAF Leeming for RTP January 2018	138, 294
ZD742	090	GR4(T)	To RAF Leeming for RTP March 2017	275
ZD743	091	GR4(T)	Written off on 3 July 2012 after colliding with ZD812 over the Moray Firth. Both crew members were killed, having failed to eject.	85
ZD744	092	GR4	WFU March 2019	189
ZD745	093	GR4	To RAF Leeming for RTP June 2016	261, 262
ZD746	094	GR4	To RAF Leeming for RTP February 2013	136, 214
ZD747	095	GR4	To RAF Leeming for RTP June 2016	126, 262
ZD748	096	GR4	To RAF Leeming for RTP January 2016	166
ZD749	097	GR4	To RAF Leeming for RTP August 2015	226

Serial	Code		Remarks	Page
ZD788	098	GR4	To RAF Leeming for RTP September 2015	276
ZD790	099	GR4	To RAF Leeming for RTP May 2016	254
ZD792	100	GR4	WFU August 2018	172
ZD793	101	GR4	WFU, GI RAF Cosford	221
ZD810	102	GR4	To RAF Leeming for RTP September 2013	260
ZD811	103	GR4	To RAF Leeming for RTP February 2013	141
ZD812	104	GR4(T)	Written off on 3 July 2012 after colliding with ZD743 over the Moray Firth. Both crew members ejected but the pilot sadly died in hospital of his injuries, while the Weapons System Officer survived with severe injuries.	224
ZD842	105	GR4(T)	To RAF Leeming for RTP March 2016	263
ZD843	106	GR4	To RAF Leeming for RTP January 2016	198
ZD844	107	GR4	To RAF Leeming for RTP May 2015	212
ZD847	108	GR4	To RAF Leeming for RTP February 2013	140
ZD848	109	GR4	WFU March 2019	196, 289
ZD849	110	GR4	WFU July 2018, to RAF Cosford for GI	228, 229, 265
ZD850	111	GR4	To RAF Leeming for RTP September 2012	245
ZD851	112	GR4	To RAF Leeming for RTP November 2016	244
ZD890	113	GR4	To RAF Leeming for RTP May 2017	151
ZD892	114	GR4	Spent most of its life as a GR.4 in storage.	No image
ZD895	115	GR4	To RAF Leeming for RTP June 2014	108
ZD996	117	GR4A	To RAF Leeming for RTP July 2012	No image
ZE116	116	GR4A	To RAF Leeming for RTP November 2013	119
ZG705	118	GR4A	To RAF Leeming for RTP May 2017	137
ZG707	119	GR4A	To RAF Leeming for RTP October 2017	236
ZG709	120	GR4A	To RAF Leeming for RTP December 2013	146, 223
ZG710		GR4A	Written off on 23 March 2003, prior to numbered codes, after being shot down by friendly fire while returning to their base in northern Kuwait, following a successful mission over Iraq. Both crew members were killed.	No image

Serial	Code		Remarks	Page
ZG711	121	GR4A	Written off on 24 October 2006 after crashing into mudflats in the Wash following a bird strike during a bombing run on Holbeach range. Both crew members ejected safely.	92
ZG712	122	GR4A	To RAF Leeming for RTP September 2013	98
ZG713	123	GR4A	To RAF Leeming for RTP April 2013	83, 102
ZG714	124	GR4A	To RAF Leeming for RTP November 2016	124
ZG726	125	GR4A	To RAF Leeming for RTP June 2012	83
ZG727	126	GR4A	To RAF Leeming for RTP June 2014	82, 242
ZG729	127	GR4A	To RAF Leeming for RTP November 2015	133
ZG750	128	GR4(T)	To RAF Leeming for RTP July 2017	281, 284, 285, 291, 292
ZG752	129	GR4(T)	WFU March 2019	271
ZG754	130	GR4(T)	To RAF Leeming for RTP January 2014	106
ZG756	131	GR4(T)	To RAF Leeming for RTP January 2014	112
ZG769	132	GR4(T)	Spent most of its life as a GR.4 in storage. To RAF Leeming for RTP July 2012	No image
ZG771	133	GR4(T)	WFU March 2019, preserved Ulster Aviation Society, Lisburn, Northern Ireland	102, 298, 299
ZG773		GR4	BAE Systems development aircraft. To RAF Leeming for RTP during October 2017.	202
ZG775	134	GR4	WFU March 2019	121
ZG777	135	GR4	To RAF Leeming for RTP November 2017	255
ZG779	136	GR4	To RAF Leeming for RTP November 2017	209, 290
ZG791	137	GR4	WFU March 2019	142
ZG792	138	GR4	Written off on 27 January 2011 after crashing into the sea near Gairloch, Scotland, with an engine fire. Both crewmembers ejected safely.	124
ZG794	139	GR4	To RAF Leeming for RTP	No image

Other than these 135 GR.4s, there are also images depicting a further 129 of the other variants of the IDS and ECR and the RAF's GR.1 and F.3, which combined with the GR.4s leads to 232 individual airframes featured throughout this book. These are listed in the tables on the next page. I could have added at least another twenty F.3s alone, but I had to draw the line somewhere.

GR.1		IDS			ECR			F.3	
Serial	Page	Serial	Air Force	Page	Serial	Air Force	Page	Serial	Page
ZA320	76	4301	GAF	71	4624	GAF	249	ZE154	65
ZA321	69	4302	GAF	75	4633	GAF	237	ZE164	56, 68
ZA326	44, 110, 113	4307	GAF	77	4646	GAF	248, 250	ZE167	67
ZA330	61, 64	4335	GAF	39	4649	GAF	251, 304, 305	ZE168	129, 195
ZA353	43	4379	GN	60	MM7051	ItAF	309, 311	ZE200	81, 91, 206
ZA355	41	4399	GAF	43				ZE201	126, 225
ZA359	46	4434	GAF	43				ZE204	65, 162
ZA360	70	MM7014	ItAF	308, 310				ZE254	175, 176
ZA361	40	MM7067	ItAF	306, 307				ZE287	89
ZA370	68	MM55000	ItAF	78				ZE288	206
ZA371	59	MM55001	ItAF	45				ZE290	45
ZA372	56	6611	RSAF	186				ZE295	92
ZA374	61	7506	RSAF	72				ZE342	130
ZA376	49	8306	RSAF	260				ZE728	183
ZA402	74	ZK113	RSAF	161, 187, 203				ZE734	41, 206, 211
ZA411	78							ZE735	169
ZA450	Rear cover							ZE737	150
ZA457	54							ZE755	155, 181
ZA461	39							ZE757	199
ZA465	50							ZE758	53
ZA469	46							ZE763	216
ZA473	52, 58							ZE764	177, 192, 194

GR.1		IDS			ECR			F.3	
Serial	Page	Serial	Air Force	Page	Serial	Air Force	Page	Serial	Page
ZA474	62							ZE785	148
ZA540	47							ZE786	158
ZA546	53							ZE790	205
ZA549	49							ZE791	84, 234
ZA556	48							ZE794	172
ZA557	44							ZE832	46
ZA559	45, 57							ZE838	55
ZA560	63							ZE839	106
ZA592	74							ZE887	192, 193, 194
ZA602	53							ZE889	68
ZA608	75							ZE934	64
ZA609	Front cover							ZE961	178
ZA613	75							ZE964	184
ZD713	51							ZE965	218, 219
ZD720	51, 73							ZE966	62
ZD740	39							ZE968	160
ZD743	73, 74, 77							ZE982	90, 180
ZD746	70							ZE983	217
ZD789	63							ZH552	94
ZD811	41							ZH553	103, 109
ZD812	52, 76							ZH554	235
ZD845	42							ZH555	157
ZD996	66							ZH557	67, 88
ZG705	72							ZH559	185

GR.1		IDS			ECR			F.3	
Serial	Page	Serial	Air Force	Page	Serial	Air Force	Page	Serial	Page
ZG707	55							ZG728	59
ZG708	50							ZG730	58
ZG711	58							ZG731	114
ZG752	66							ZG753	153
ZG769	54							ZG757	188
ZG771	67							ZG778	61
ZG791	62							ZG780	154
ZG794	72							ZG793	132
								ZG797	127

Above left: ZA461/GA was the first GR.1 delivered to 20 Squadron in March 1984, although the squadron didn't officially stand up at RAF Laarbruch, West Germany, until 29 June 1984. It is pictured here at low-level over West Germany during 1986. *(Tony Paxton)*

Above right: IDS(T) 4335/G-38 of the TTTE, belonging to what was then the West German Air Force. It is seen here over Yorkshire on 1 September 1986 during a low-level sortie from its base at RAF Cottesmore. *(Tony Paxton)*

Right: Proudly displaying its camouflage colour scheme and large 31 Squadron markings, GR.1 ZD740/DA, is seen at low-level over the northern plains of West Germany, during a low-level flight from its home base of RAF Brüggen. Circa 1987. *(Tony Paxton)*

Left: The earliest image in the book taken from the side of a valley sees Tornado GR.1 ZA361/B-57, of the TTTE based at RAF Cottesmore, pictured over the Thirlmere reservoir in the Lake District during a training sortie on 28 June 1988. *(Adrian Walker)*

Opposite above: F.3 ZE734/CJ wearing the original V(AC) Squadron markings applied to F.3s and based at RAF Coningsby, seen at low-level over the sea during 1988. *(Tony Paxton)*

Opposite below left: Another TTTE jet, this time GR.1 ZA355/B-54, seen as it approaches Dunmail Raise in the Lake District on 16 August 1988. Dunmail Raise, often shortened to just Dunmail, is on the route between Windermere and Bassenthwaite within LFA17 and was a favourite route for Tornado crews from all bases, including the RAF Germany based squadrons. *(Adrian Walker)*

Opposite below right: RAF Germany Tornado GR.1, ZD811/DF, based at RAF Brüggen with 31 Squadron, pictured here passing through Dunmail Raise on 1 November 1988. Note the additional external fuel tank under the fuselage, a common configuration for RAF Germany based Tornados, but not something that was exclusive to them, with aircraft from RAF Marham also being quite common. *(Adrian Walker)*

The Panavia Tornado at Low-Level • 41

14 Squadron GR.1, ZD845/BA, based at RAF Brüggen in West Germany, seen passing through Dunmail Raise on 7 March 1989. During the late 1980s several RAF Germany based Tornado GR.1s received coloured tails for a low-level visibility trial, with ZD845 receiving a red tail for this purpose. This aircraft crashed on 26 February 1996 while on an air test from RAF Brüggen while operated by IX(B) Squadron. After a mechanical fault caused a fire warning, the crew attempted to divert to RAF Laarbruch, but a control loss on approach led to the crew ejecting and the aircraft crashing eight miles from the base. *(Adrian Walker)*

GR.1 ZA353/B-53 of the TTTE, seen passing through Dunmail Raise on 25 April 1989. The TTTE were very common within LFA17, particularly the route through Thirlmere, the introduction of the Tornado GR.1 and basing of the TTTE at RAF Cottesmore leading to a review of the UKLFS in 1979, which opened up the previously restricted area of the Lake District to low-level flying. *(Adrian Walker)*

4434 was one of a pair of West German Air Force IDS Tornados of JBG 31, seen routing north through Dunmail Raise on 10 May 1989. Four JBG 31 IDSs were deployed to RAF Coningsby for a squadron exchange with 29 Squadron and their Tornado F.3s. *(Adrian Walker)*

Following on from the previous image, the afternoon of 10 May 1989 saw a return of the JBG 31 Tornado IDSs of the West German Air Force through Dunmail Raise. 4399 is pictured as one of the afternoon pair, however on this occasion the pair were noted in the company of a 27 Squadron Tornado GR.1 from RAF Marham. *(Adrian Walker)*

The Panavia Tornado at Low-Level • 43

Left: GR.1 ZA557/JG of 27 Squadron based at RAF Marham, seen passing through Dunmail Raise on 26 May 1989. My personal favourite Tornado squadron, the red band blending well with the green/grey camouflage scheme, although unfortunately it was one of the first RAF Tornado squadrons to disband when they effectively became 12(B) Squadron, converting to the GR.1B in late 1993 to replace the Buccaneer in the maritime attack role, in the process relocating to RAF Lossiemouth in early 1994. *(Adrian Walker)*

Below: The Royal Aircraft Establishment (RAE) operated a number of red, white and blue liveried aircraft types from their bases at Farnborough and Bedford. Affectionately known as 'Raspberry Ripple', many of these aircraft could be seen within the UKLFS, one of which was Tornado GR.1P, ZA326, which operated from Thurleigh airfield, Bedford. It was seen here passing Dunmail Raise on 26 May 1989. *(Adrian Walker)*

GR.1 ZA559/L of 617 Squadron based at RAF Marham, pictured at Dunmail Raise on 7 June 1989. One of only two UK based frontline RAF Tornado squadrons at the time, alongside 27 Squadron also based at RAF Marham, they were at this time carrying a single letter tail code. A few years later they would revert back to using the AJ-* codes worn originally by the squadron when formed in 1943 with Lancaster bombers for 'Operation Chastise', the Dambuster raids over Germany for which they are famous. *(Adrian Walker)*

Italian Air Force IDS(T), MM55001/I-40, passing through Dunmail Raise on 18 July 1989. This was another jet from the TTTE based at RAF Cottesmore, responsible for training Tornado crews of the three partner nations of the UK, West Germany and Italy. This aircraft crashed during a training flight in Italy on 6 June 2007, with both crewmembers ejecting safely. *(Adrian Walker)*

The Operational Conversion Units (OCUs) were responsible for training crews on types prior to being allocated to frontline squadrons. For the Tornado F.3 this was 229 OCU, which had the shadow reserve designation of 65(R) Squadron. F.3(T) ZE290/AD belonging to 229 OCU/65(R) Squadron, is pictured passing through Dunmail Raise on 4 July 1989 during a low-level training flight from their home base of RAF Coningsby. *(Adrian Walker)*

Above left: A three-ship of 20 Squadron GR.1s from RAF Laarbruch, West Germany, led by ZA469/GD and seen at low-level over the North York Moors, while inbound to RAF Leeming after a low-level sortie utilising LFA17. *(Ian Black)*

Above right: Another TTTE image, this one showing GR.1, ZA359/B-55 passing through a snowy Dunmail Raise on 16 February 1990. Dunmail Raise will be mentioned many times in this book as one of the best known photography locations within the Lake District and Adrian Walker's personal favourite for its consistency. A large pile of rocks lie between the two carriageways of the A591 road at the top of the pass and local tradition suggests that this is the burial place of King Dunmail, the last king of Cumberland. *(Adrian Walker)*

Left: 23 Squadron F.3 ZE832/EB, based at RAF Leeming, is seen at low-level over the Möhne dam, West Germany, during January 1990. The Möhne dam was one of the dams destroyed by 617 Squadron's Lancaster bombers during Operation Chastise on 16 May 1943. *(Tony Paxton)*

GR.1(T) ZA540/JQ, one of a pair of 27 Squadron Tornados seen passing through Dunmail Raise on 23 May 1990. This aircraft was written off on 1 September 1991 after crashing into the Bristol Channel when all control was lost. Both crew members ejected safely. *(Adrian Walker)*

During 1990 the TWCU/45(R) Squadron based at RAF Honington, operated three specially marked Tornado GR.1s as display aircraft for the airshow season. The one pictured here is thought to be ZA556, one of a pair seen passing through Dunmail Raise on 24 May 1990. *(Adrian Walker)*

The other two TWCU/45(R) Squadron display aircraft were ZA543 and ZA545. This image, also shot at Dunmail Raise just a day later on 25 May 1990, shows this GR.1 carrying external fuel tanks unlike the previous image. While it's entirely possible that the tanks had been fitted between flights, it is more likely that this image portrays one of the other airframes mentioned above, and also of note is that this aircraft is carrying 1,000lb inert bombs. As a side note, ZA545 was written off just a few weeks later on 14 August 1990 with the loss of both crewmembers, after colliding with ZA464 at low-level over the North Sea, 10nm north east of Spurn Head, Yorkshire. The pilot of ZA464, the lead aircraft of a pair of 20 Squadron Tornado GR.1s from RAF Laarbruch, West Germany, was the only survivor. *(Adrian Walker)*

ZA376/JL was one of a pair of 27 Squadron GR.1s to pass through Dunmail Raise at 11.52 on 31 May 1990. This aircraft was written off on 10 May 1991 after a control loss led to it crashing near Lübberstedt, Germany, while on loan to 20 Squadron, but being flown by a XV Squadron crew, both of whom ejected. *(Adrian Walker)*

GR.1(T) ZA549 of XV Squadron based at RAF Laarbruch, West Germany, seen at low-level over North Yorkshire during a low-level training flight. This aircraft was carrying markings for the squadrons 75th anniversary and had a certain Flight Lieutenant John Nichol in the back seat, this image being taken a few months before the start of the Gulf War that would have a major impact on the navigator's life. *(Ian Black)*

One of the most famous Tornado GR.1s was ZA465/FK of 16 Squadron, based at RAF Laarbruch, Germany. It was given a sharks mouth and 'Foxy Killer' nose art for Operation Granby during the Gulf War and flew the most bombing missions of any RAF Tornado in the war. It went on to become a GR.1B with 12(B) Squadron, before being retired to the Imperial War Museum at Duxford. It is pictured here at the north end of Thirlmere, still wearing its desert pink Alkaline Removable Temporary Finish (ARTF) camouflage, on 22 April 1991. *(Adrian Walker)*

GR.1A ZG708/C of XIII Squadron based at RAF Honington, photographed at Dunmail Raise on 10 July 1991. XIII Squadron were the third UK based frontline squadron when they reformed on the Tornado on 1 January 1990 and one of two squadrons to operate the GR.1A variant in the reconnaissance role, alongside II(AC) Squadron. This particular aircraft was written off on 1 September 1994 while flying down Glen Ogle, Perthshire, Scotland, on a low-level training flight from their then home base of RAF Marham. The aircraft flew into the ground after a sudden control movement with reheat applied, killing both crewmembers. The cause was never identified. *(Adrian Walker)*

Above: Another Gulf War veteran still wearing its ARTF camouflage was ZD720/CK of 17 Squadron, also seen at Dunmail Raise on 10 July 1991, around an hour and ten minutes behind the XIII Squadron jet in the previous image. *(Adrian Walker)*

Right: Also at Dunmail Raise on 10 July 1991, around an hour after the Granby jet in the previous image, is GR.1(T) ZD713 of the TWCU/45(R) Squadron based at RAF Honington. *(Adrian Walker)*

ZD812/FV and ZA473/FM, a pair of 16 Squadron Tornado GR.1s from RAF Laarbruch, Germany, seen at low-level over North Yorkshire. 16 Squadron were one of the lead Tornado squadrons in the Operation Granby deployment for the Gulf War, along with 20 Squadron, deploying JP233 and 1,000lb bombs during low-level missions against Iraqi targets. In this image during 1991, they were in their final weeks of operations before disbanding, which occurred on 11 September 1991. The squadron would reform on 1 November 1991 at RAF Lossiemouth, as a Reserve (R) Squadron for 226 OCU, the Jaguar Operational Conversion Unit. *(Ian Black)*

An unidentified TTTE IDS belonging to the German Air Force, passes the north end of Thirlmere in the Lake District on 5 November 1991. *(Adrian Walker)*

Another scenic image from the north end of Thirlmere sees GR.1(T) ZA602/AZ of IX(B) Squadron based at RAF Brüggen in Germany, passing through on 23 January 1992. This area is known as Smaithwaite and is still one of the most popular areas in the Lake District for low-level photographers today. *(Adrian Walker)*

Smaithwaite again and ZA546/JB of 27 Squadron based at RAF Marham is pictured taking the turn towards Keswick to then go on to Bassenthwaite on 20 May 1992. 27 Squadron would disband in September 1993 with their aircraft being converted to GR.1B standard and transferred to 12(B) Squadron to replace the Buccaneer in the maritime attack role, eventually moving to RAF Lossiemouth in March 1994. *(Adrian Walker)*

A picturesque view of F.3, ZE758/C, of 1435 Flt, as it cruises around the Falkland Islands at low-level with wings fully swept. Tornado F.3s replaced F-4 Phantoms at RAF Mount Pleasant during 1992, providing air defence for the Falkland Islands, South Georgia and the South Sandwich Islands. Historically, the four aircraft have carried the tail codes of F, H, C and D, referring to Faith, Hope, Charity and Desperation and also carry the Maltese cross for its connection to Malta where they were first formed during July 1941. *(Ian Black)*

Above: Back to Adrian Walker's favourite haunt of Dunmail Raise and GR.1(T) ZG769/AY of IX(B) Squadron heads away from Dunmail on 19 May 1994. ZG769 went on to be converted to GR.4 standard, but is one of ten GR.4s that I haven't been able to source a photo of for the book, due to it being an early candidate for storage. *(Adrian Walker)*

Left: GR.1B ZA457/AJ-J of 617 Squadron, passing Dunmail on 22 June 1994, the squadron having now reverted to using the original AJ-* codes. After being based at RAF Marham for many years, the famous 'Dambusters' Squadron of 617 had by now relocated to RAF Lossiemouth with the GR.1B variant in the maritime attack role, alongside 12(B) Squadron. *(Adrian Walker)*

F.3 ZE838/GH of 43(F) Squadron, one of the two F.3 squadrons originally based at RAF Leuchars alongside 111(F) Squadron. Known as the 'Fighting Cocks' with a Gamecock as their squadron badge, 43(F) Squadron generally seemed to avoid me in the valleys, but Adrian Walker caught this one passing through Dunmail Raise on 1 September 1994. *(Adrian Walker)*

GR.1A ZG707/B of XIII Squadron, passing through Dunmail Raise on 21 December 1994. The RAF introduced thirty of the GR.1A variant as a dedicated reconnaissance platform during the late 1980s, with fourteen converted from standard GR.1s and sixteen new builds. Originally the majority of converted aircraft populated II(AC) Squadron at RAF Laarbruch, while the new builds were given to XIII Squadron at RAF Honington. Both squadrons would be based at RAF Marham by the start of February 1994. *(Adrian Walker)*

Also at Dunmail Raise on 21 December 1994 was GR.1A ZA372/E of II(AC) Squadron, just over an hour behind the XIII Squadron jet in the previous image. This was one of the fourteen converted GR.1s which came from the batch between ZA369 and ZA406. *(Adrian Walker)*

A snowy Dunmail Raise provides the backdrop for Tornado F.3 ZE164/DA on 9 March 1995. Largely unmarked other than the XI(F) Squadron code, this aircraft was written off after a bird strike on 6 January 2011 while operating with 111(F) Squadron. Although landing safely at RAF Leuchars and being deemed repairable, the imminent withdrawal of the F.3 from RAF service meant the repair was not authorised and it was taken by road to RAF Leeming for scrap. *(Adrian Walker)*

GR.1 ZA559/F of XV(R) Squadron, seen wearing the squadron's 'MacRobert's Reply' special scheme carried by the aircraft coded 'F', with added 80th Anniversary years on the tail fin. Having disbanded on Tornados at RAF Laarbruch at the end of 1991, XV Squadron were given reserve status with the TWCU on 1 April 1992, replacing 45(R). In November 1993 the squadron would relocate from RAF Honington to RAF Lossiemouth, where it was likely en route to in this image, as it passes through Dunmail Raise on 12 April 1995. The eventual demise of the TTTE would see all UK Tornado training shifted to XV(R) Squadron. *(Adrian Walker)*

Above left: F.3 ZG730/CC of V(AC) Squadron based at RAF Coningsby, photographed from the eastern side of Dunmail Raise on 25 April 1995, one of six images from this date. RAF Coningsby housed three Tornado F.3 squadrons, with 29 Squadron and the OCU joining V(AC) Squadron. *(Adrian Walker)*

Above right: 1995 witnessed the 80th Anniversary of several RAF squadrons, as seen two images previous with XV(R) Squadron. This image depicts GR.1A ZG711 from RAF Marham, wearing the XIII Squadron 80th Anniversary markings as it passes through Dunmail Raise on 25 April 1995, about an hour after the F.3 in the previous image. *(Adrian Walker)*

Left: Another image from 25 April 1995 sees GR.1B, ZA473/FG of 12(B) Squadron based at RAF Lossiemouth, as it passes through Dunmail Raise less than thirty minutes after the XIII Squadron anniversary jet in the previous image. *(Adrian Walker)*

Into the afternoon of 25 April 1995 and one of a pair of V(AC) Squadron F.3s, ZG728/CI, is seen passing through Dunmail Raise. This was a superb day for movements with twenty-two aircraft through between 09.53 and 14.02, fourteen of which were Tornados.
(Adrian Walker)

25 April 1995 again and the next image from Dunmail Raise shows what an excellent day it was for Tornados, as II(AC) Squadron GR.1A ZA371/C from RAF Marham, is seen with a white ARTF finish over the usual green to create an arctic camouflage for its time deployed to Norway on exercise earlier in the year.
(Adrian Walker)

The final of the six images from 25 April 1995 sees IDS 4379/G-76, of the TTTE's A-Squadron. Germany had the largest contingent of aircraft within the TTTE with twenty-three aircraft, the UK providing ninetten aircraft and Italy six. This particular aircraft was wearing German Navy camouflage. *(Adrian Walker)*

Above left: GR.1B ZA374/AJ-D of 617 Squadron based at RAF Lossiemouth, seen at Dunmail Raise on 1 May 1995. This Gulf War veteran now resides at the National Museum of the United States Air Force, at Wright-Patterson Air Force Base, Ohio. *(Adrian Walker)*

Above right: Another TTTE, A-Squadron Tornado, this time in the form of RAF GR.1(T) ZA330/B-08, passing through Dunmail Raise at 09.52 on 18 May 1995. This aircraft was written off after a mid-air collision with Cessna 152 G-BPZX, near Mattersey, Nottinghamshire, as it departed RAF Cottesmore into the low-level portion of a training sortie on 21 January 1999, killing both of the Tornado crewmembers and the pilot and passenger of the Cessna. The Cessna crashed at the point of impact while the Tornado came down 3 km away. Even more tragic was that this was the final conversion course before the closure of the TTTE. *(Adrian Walker)*

Right: Another RAF Coningsby-based F.3 at Dunmail Raise, this time ZG778/BG, belonging to 29 Squadron. The only fully marked 29 Squadron F.3 I had available for this book, it was seen here on 28 June 1995. *(Adrian Walker)*

Above left: Another of the RAF 80th Anniversary Tornados was GR.1 ZG791, which carried the anniversary scheme for 31 Squadron who were known as the 'Goldstars'. It is pictured here at Dunmail Raise on 13 July 1995. *(Adrian Walker)*

Above right: GR.1B ZA474/FF of 12(B) Squadron based at RAF Lossiemouth, passes through Dunmail Raise on 31 August 1995. *(Adrian Walker)*

Left: F.3(T) ZE966/DZ of XI(F) Squadron based at RAF Leeming, passes through Dunmail Raise on 17 August 1995. Historically, the two-digit codes worn by RAF frontline squadrons that contained the second letter T, W, X, Y, or Z as in this image, generally indicated that the aircraft was a twin stick trainer. One exception to this was II(AC) Squadron, which used the squadron nickname Shiny Two to provide codes to spell SHINEY TWO AC. *(Adrian Walker)*

RAF Germany GR.1, ZD789/AM of IX(B) Squadron based at RAF Brüggen, seen at Dunmail Raise on 23 August 1995. This aircraft was written off on 23 February 1996 during an emergency landing at RAF Brüggen, with the aircraft on fire. The aircraft suffered extensive damage and was withdrawn from use, moved to RAF Shawbury and scrapped in April 2002. *(Adrian Walker)*

Also passing through Dunmail Raise on 23 August 1995, thirty-three minutes behind ZD789 in the previous image, was GR.1 ZA560 of the TTTE. This aircraft was seen wearing an all-over black colour scheme as a display jet for the 1995 airshow season. *(Adrian Walker)*

Above: F.3(T) ZE934/DX of XI(F) Squadron, seen passing through Dunmail Raise on 1 March 1996. This aircraft now resides at the National Museum of Flight, East Fortune, Scotland. *(Adrian Walker)*

Left: GR.1(T) ZA330/B-08 of the TTTE, seen taking a high line through Dunmail Raise on 1 March 1996, revealing a snowy backdrop not present in the previous image of ZE934 on the same date. This is the first image in the book showing a Tornado GR.1 in the all-over grey colour scheme that would eventually adorn all RAF Tornado GRs, as well as, sadly, the IDSs of the German, Italian and Saudi Air Forces, although the latter would take many more years before losing their original sand camouflage. *(Adrian Walker)*

F.3(T) ZE154/GI of 43(F) Squadron, based at RAF Leuchars, passing Dunmail Raise on 27 March 1996. *(Adrian Walker)*

F.3 ZE204/DD of XI(F) Squadron, carrying a squadron code without squadron markings, pictured at Dunmail Raise on 29 March 1996 as one of a four-ship of XI(F) Squadron aircraft. *(Adrian Walker)*

GR.1A ZD996/I of II(AC) Squadron based at RAF Marham, seen at Dunmail Raise on 10 May 1996. This aircraft was the first new build GR.1A and went on to be upgraded to a GR.4A, but is one of only ten GR.4s that I've been unable to source a low-level image of. *(Adrian Walker)*

GR.1(T) ZG752/XIII, one of XIII Squadron's twin stick trainers, seen passing through Dunmail Raise on 6 June 1996. The two GR.1A squadrons based at RAF Marham were regular users of LFA17.
(Adrian Walker)

Right: F.3 ZE167/HX of 111(F) Squadron based at RAF Leuchars, pictured passing Dunmail Raise while on a low-level training sortie through LFA17 on 13 June 1996. *(Adrian Walker)*

Below left: The 'Goldstars' of 31 Squadron were regular visitors to the Lake District during their time with RAF Germany, based at RAF Brüggen. GR.1(T) ZG771/DW was photographed at Dunmail Raise on 25 July 1996. *(Adrian Walker)*

Below right: F.3(T) ZH557/CT of V(AC) Squadron, through Dunmail Raise on 21 October 1996. *(Adrian Walker)*

Left: F.3 ZE889 of the Tornado F.3 Operational Evaluation Unit (OEU) based at RAF Coningsby, passing through Dunmail Raise on 11 November 1996. For many years the Tornado F.3 OEU operated out of RAF Coningsby while the Strike Attack OEU operated the Tornado GR, Jaguar and Harrier out of Boscombe Down. They would merge in 2004 becoming the Fast Jet and Weapons OEU (FJWOEU) and be based at RAF Coningsby. *(Adrian Walker)*

Above: GR.1A ZA370 of II(AC) Squadron based at RAF Marham, seen at Dunmail Raise on 14 February 1997. One of the first operational GR.1As, this aircraft still sports a large number of mission marks from its time in the Gulf War as part of Operation Granby, when it was one of six GR.1As to deploy with II(AC) Squadron. *(Adrian Walker)*

Left: F.3 ZE164/DA at Dunmail Raise on 6 March 1997. This aircraft was pictured earlier in the book with a snowy background on 9 March 1995, but is now seen wearing the markings of XXV(F) Squadron but with an XI(F) Squadron code, almost exactly two years later. Mixed markings were something that could be quite common at times with RAF aircraft. *(Adrian Walker)*

The RAF Tornado GR.1 airshow demonstration was provided by a number of squadrons over the years, with the TWCU, TTTE, XV(R) all carrying the baton at some point, with even a frontline squadron in the shape of 27 Squadron being utilised for a short time. This image shows GR.1 ZA321/B-58 of the TTTE, wearing the still relatively new all-over grey scheme as it passes through Dunmail Raise on 9 April 1997. This was one of the display jets with the TTTE for the 1996 and 1997 display seasons. *(Adrian Walker)*

Above: GR.1 ZD746/AB of IX(B) Squadron based at RAF Brüggen, Germany, seen passing through Dunmail Raise on 10 April 1997. *(Adrian Walker)*

Left: GR.1 ZA360/B-56, another of the TTTE's grey display aircraft, seen at Dunmail Raise on 15 April 1997. *(Adrian Walker)*

Another German Air Force IDS(T), 4301/G-20 of the TTTE's A Squadron, looking resplendent in its camouflage as it passes through Dunmail Raise on 15 April 1997, some twenty minutes behind ZA360 in the previous image. *(Adrian Walker)*

Left: 15 April 1997 again and just six minutes behind 4301 in the previous image was GR.1A ZG705/J of XIII Squadron. This is the first image in the book of a frontline squadron in the new grey scheme. *(Adrian Walker)*

Above: Royal Saudi Air Force Tornado IDS 7506, complete with UK serial, ZH923, captured passing through Dunmail Raise on 10 July 1997. Aircraft built in the UK for export are allocated UK serials for flight testing and removed upon delivery. This aircraft was flying from the British Aerospace facility at Warton in Lancashire where it was built and would be delivered to the RSAF on 24 July 1997. *(Adrian Walker)*

Left: GR.1 ZG794/BP of 14 Squadron, based at RAF Brüggen in Germany, passing Dunmail Raise on 18 August 1997. This was another GR.1 that would be upgraded to GR.4 standard, but one of the ten GR.4s that I couldn't source a low-level image of. *(Adrian Walker)*

GR.1 ZD720/AG of IX(B) Squadron, based at RAF Brüggen, Germany, seen passing Dunmail Raise on 21 October 1997. This was the third of four Tornado GR.1s to pass through at 15.39 on this day, behind ZA355 of the TTTE and another IX(B) Squadron jet coded AX.
(Adrian Walker)

The fourth aircraft of the four through Dunmail Raise at 15.39 on 21 October 1997 was GR.1(T) ZD743/CX of 17(F) Squadron, also based at RAF Brüggen.
(Adrian Walker)

Above left: Another view of ZD743/CX of 17(F) Squadron on 21 October 1997, as it catches the best of the autumnal light passing through Dunmail Raise. The three main routes through LFA17 are all flowed south to north, making it frustrating at times for photography. Chasing the best light requires a change of sides as the sun moves around, with Dunmail Raise in particular, making that point very well. Depending on the time of year, the best of the light is from the western side from mid-afternoon, with the evening light in the summer months being exceptional. *(Adrian Walker)*

Top right: GR.1 ZA592/B, carrying partial 617 Squadron markings but with single digit code that had been superseded by AJ-* codes a few years earlier. It had been in a period of storage and so had possibly been brought back into service recently and was yet to receive markings for whichever squadron it was now allocated to. It was seen at Dunmail Raise on 29 October 1997. *(Adrian Walker)*

Above right: GR.1 ZA402, passing through Dunmail Raise on 29 October 1997. This Tornado spent almost all of its life as a trials and development aircraft, spending time at both Warton and Boscombe Down, seen here while based at the latter with the Aircraft and Armament Evaluation Establishment (A&AEE), but carrying tail markings for the TWCU. It was used as a trials aircraft for the GR.1A development, the small brown window below the cockpit showing evidence of the Sideways Looking Infra Red (SLIR) equipment having been installed, although it was never officially classified as a GR.1A. It also went on to become one of three GR.4 development aircraft flying from BAE Warton, when it did however become classified as a GR.4A. *(Adrian Walker)*

Right: German Air Force IDS(T), 4302/G-21 of the TTTE, passing through Dunmail Raise on 1 December 1997. *(Adrian Walker)*

Below left: A sign of things to come. On 19 March 1998, RAF GR.1 ZA608, is seen approaching Dunmail Raise in the grey scheme and completely devoid of squadron markings, something that would start to become commonplace, although thankfully this was several years before it would become very widespread. ZA608 was another GR.1 that would be upgraded to GR.4 standard and another of the ten GR.4s that have alluded my efforts to trace a low-level image of. *(Adrian Walker)*

Below right: XV(R) Squadron GR.1 ZA613/TL, based at RAF Lossiemouth, passing through Dunmail Raise on 19 March 1998. *(Adrian Walker)*

GR.1(T) ZD812/BW of 14 Squadron based at RAF Brüggen, Germany, approaches Dunmail Raise on 19 March 1998. This image perfectly demonstrates how aircraft become silhouetted as the sun moves around the south, necessitating a move to the other side of the valley for the afternoon. *(Adrian Walker)*

This image sees one of the earliest RAF GR.1(T)s in the form of ZD320/B-01 of the TTTE. This final image from 19 March 1998 sees it pass through Dunmail Raise during the afternoon, showing it passing from right to left, unlike left to right in the previous three images that were taken from the eastern side during the morning. This aircraft now resides at RAF Cosford and has been superbly restored and painted in original camouflage scheme with 27 Squadron markings adorning the right side of the aircraft and 17(F) Squadron markings on the left, at the time of writing. *(Adrian Walker)*

Above: GR.1 ZD743/CX of 17(F) Squadron based at RAF Brüggen, Germany, passing through Dunmail Raise on 31 March 1998.
(Adrian Walker)

Right: German Air Force IDS(T) 4307/G-26, of the TTTE's A Squadron, making a high pass through Dunmail Raise on 16 July 1998.
(Adrian Walker)

Above: Italian Air Force IDS(T) MM55000/I-42, of the TTTE's B Squadron, passing Dunmail Raise on 10 March 1999. *(Adrian Walker)*

Left: GR.1(T) ZA411/AJ-S of 617 Squadron, making the most of the Spring sunshine as it passes through Dunmail Raise on 5 April 2000. *(Adrian Walker)*

The first of my own images and my very first day doing low-level photography at the Bwlch area of the Mach Loop. I have been unable to find a logbook entry for this day, so all I know is that it was June 2001. It shows an unidentified GR.4A in XIII Squadron markings, possibly not long after upgrade to GR.4 standard. *(Scott Rathbone)*

Back into LFA17 and my very first trip to Cumbria for low-level photography was May 2002, again though I'm unable to find the exact date. My first movements through there were a pair of XIII Squadron Tornado GR.4As, one of which was this unidentified example. *(Scott Rathbone)*

These three images show a pair of XI(F) Squadron F.3s from RAF Leeming, entering and exiting the M6 Pass during September 2002, with one passing through opposite the photo location. F.3s were relatively common through the M6 Pass and could generally be seen morning and afternoon on practice Combat Air Patrols (CAPs) over the top of the Pass and the surrounding area within the Overland Training Area (OTA), looking for any unsuspecting traffic in the LFA to 'bounce'. By 2007 this practice was starting to diminish as traffic levels within the LFAs reduced, although in my experience the only area that regularly saw this kind of activity was the M6 Pass and areas to the east, with the Lake District barely witnessing any CAP or bouncing activity. *(Scott Rathbone)*

Above: F.3 ZE00/UM, still wearing the markings of V(AC) Squadron which had disbanded on 30 September 2002, is seen in the M6 Pass on 28 October 2002 carrying a converted BOZ Pod visible underneath the end of the left wing. BOZ Pods were counter-measures pods generally referred to as chaff and flare dispensers, and fitted to the Tornado IDS/GR. During 1995 some were converted to be used as a Towed Radar Decoy (TRD) and fitted to Tornado F.3s for use during the Balkans campaign. *(Chris Chambers)*

Right: Another view of F.3 ZE200/UM in the same fit providing another view of the TRD, just a week or so after the previous image and again in the M6 Pass. *(Scott Rathbone)*

Above left: What was the XV Squadron GR.1(T) 75th Anniversary aircraft during 1990, ZA549, now upgraded to GR.4(T) standard, is seen rushing through the Mach Loop on 7 January 2003. It was thought to have been on a test flight out of RAF St Athan, and considering it was an early GR.4 conversion and still wearing camouflage, it is believed that the aircraft had been in storage and was now in the process of being brought back into service. *(Chris Chambers)*

Above right: The somewhat controversial invasion of Iraq by a US led coalition during early 2003, known as the Iraq War and also referred to as Gulf War 2, saw UK involvement in the form of Operation TELIC, with British Tornados once again engaged in combat operations against Iraqi targets. Reflecting a change in tactics to a medium to high level role rather than predominately low-level, the desert pink camouflage was superseded by an all-over light grey ARTF scheme, with the first known UK low-level sighting of the scheme being GR.4A, ZG727/L of XIII Squadron, which was seen passing through a snowy M6 Pass on 5 February 2003, unusual in itself as the aircraft was still carrying a black nose rather than the grey nose that would be utilised during the operation. *(Chris Chambers)*

Opposite above: Upon their return from Operation TELIC, some of the Tornados retained their ARTF scheme for several weeks and were caught within the UKLFS. Two were noted in the Mach Loop on 3 July 2003, the first being GR.4A ZG713/G, flying with II(AC) Squadron from RAF Marham, seen with its Operation TELIC, Combat Air Wing badge on the tail. This was a special badge made up for the deployment comprising the markings of each of the four Marham based squadrons of II(AC), IX(B), XIII and 31 Squadrons and also 617 Squadron, which was the only squadron involved based at RAF Lossiemouth. The aircraft were pooled, which led to a mix of other squadrons aircraft being operated. *(Scott Rathbone)*

Opposite below left: Also in the Mach Loop on 3 July 2003 was GR.4 ZA542/DM of 31 Squadron. Several Tornados carried nose art during the deployment and this generally reflected the aircraft tail code, so in this case the code DM proved to be perfect for the cult 1980s cartoon character 'Danger Mouse', as can be seen in this image. This aircraft also passed through the Mach Loop the previous day, 2 July, but stayed high along the Tal-y-Llyn valley. *(Scott Rathbone)*

Opposite below right: Another ex-Operation TELIC GR.4A in the Mach Loop was ZG726/K of XIII Squadron, based at RAF Marham. Tornado nose art was generally painted under the cockpit on the left side of the aircraft, but unusually ZG726 had it applied to the right side, so therefore not visible in this image. It carried Kylie Minogue nose art reflecting the tail code 'K' and is seen here taking the turn at Corris corner on 14 July 2003. *(Scott Rathbone)*

Above: The last of the ARTF grey Tornados in the book (but not the last connected with Operation TELIC), sees ZA592/BJ as it approaches the north end of Thirlmere during late July 2003. Although carrying a 14 Squadron tail code from RAF Lossiemouth, it was in fact flying out of RAF Marham with one of the squadrons based there, after returning from the Middle East. This was one of a small number of Tornados to have carried sharks' mouths and this image clearly shows that along with mission marks and the Combat Air Wing badge. *(Chris Chambers)*

Left: The start of a day I will never forget, one of almost non-stop action from mid-morning until early afternoon with more traffic from mid-late afternoon. F.3(T) ZE791/XY of 111(F) Squadron based at RAF Leuchars, seen making its second pass of the day through Dunmail Raise some sixteen minutes after its first pass. This was 4 August 2003, a day mentioned in the introduction as 'Mad Monday'. *(Scott Rathbone)*

GR.4(T) ZD743/TQ of XV(R) Squadron, based at RAF Lossiemouth, passing through Dunmail Raise on 4 August 2003. Tornado GRs traditionally carried black noses, that until the advent of Operation TELIC adopting a grey nose for use with the light grey ARTF scheme, this being an early example of the grey nose being retained as the norm. This aircraft was tragically lost along with its two crewmembers on 3 July 2012 after colliding with Tornado GR.4(T) ZD812, also flying with XV(R) Squadron from RAF Lossiemouth. Both aircraft were holding for an attack on Tain Range when the collision occurred over the Moray Firth, twenty-five miles south east of Wick, with the two aircraft in opposite banked turns they were unable to see each other and struck each others underside. *(Scott Rathbone)*

The inclusion of this innocuous, backlit and unmarked GR.4 might seem a little odd when I had to leave out so many amazing images, however this is ZA491, the third GR.4 to be lost when it crashed into the North Sea off Newton Point, Northumberland, on 22 July 2004, during a training flight operating with 31 Squadron from RAF Marham, the crew ejecting safely. Due to its early demise, along with ZA599 in 2002 and ZG710 in 2003, they would be the only frontline GR.4s not to receive a numbered code, the only other GR.4s not to receive one being the development aircraft of ZA402, ZD708 and ZG773. ZA491 is pictured here at Dunmail Raise on 4 August 2003, as number three of a three-ship from RAF Marham. *(Scott Rathbone)*

With the weather being stunning all day on 4 August 2003, it would be easy to get carried away and use many images from that day. However I've tried to refrain from that and pick out noteworthy images and for me this was one of the best low-level images I had ever taken at the time, as well as being the last movement of the day, as far as I was aware. It shows GR.4(T) ZA544/FZ of 12(B) Squadron based at RAF Lossiemouth, as it speeds through Dunmail at 17.15 and was one of several 12(B) Squadron aircraft seen that day. In fact this was its second pass of the day having been one of the mass of Tornados through between 11.30 and 12.41, which included a four-ship. *(Scott Rathbone)*

F.3(T) ZH557/NT wearing XXV(F) Squadron markings based at RAF Leeming. This was the lead aircraft of a pair of twin-stick F.3s, seen passing through Thirlmere on 1 September 2003. *(Scott Rathbone)*

The second aircraft of the pair of F.3s through Thirlmere on 1 September 2003 was F.3(T) ZE287/TO, wearing dual markings for XI(F) Squadron based at RAF Leeming and 56(R) Squadron based at RAF Leuchars. These two aircraft unusually appeared from the north of Thirlmere at approximately 2,000ft, heading south above the eastern side of the valley. Upon reaching the Grasmere area they made a right turn to drop down over Grasmere and made a low-level run through the valley. This was quite possibly a pair on CAP that had ventured west at low-level in search of trade. *(Scott Rathbone)*

Left: F.3 ZE982/VV of XXV(F) Squadron based at RAF Leeming, pictured as it exits the Thirlmere valley at the north end of Thirlmere reservoir on 4 September 2003. This aircraft was destroyed with the loss of both crewmembers on 2 July 2009, while flying with 43(F) Squadron based at RAF Leuchars. Flying as 'Blacksmith 1' the lead of a pair of F.3s conducting a low-level navigation exercise as part of a larger training sortie, the aircraft impacted the ground on the north slope of Glen Kinglass, near Arrochar, Argyll and Bute, Scotland. *(Scott Rathbone)*

Below: A scenic view of an unidentified F.3 as it makes its way through the M6 Pass on 24 September 2003. *(Scott Rathbone)*

XI(F) Squadron F.3, ZE200, seen in the M6 Pass on 7 October 2003. This was one of a pair of XI(F) Squadron F.3s on CAP that afternoon that proceeded to make three passes through the valley in thirty minutes. *(Scott Rathbone)*

Above: The second of the pair of XI(F) Squadron F.3s through the M6 Pass on 7 October 2003 was ZE295/AY. This was another dual marked aircraft carrying the markings of both XI(F) and 56(R) Squadrons. *(Scott Rathbone)*

Right: GR.4A ZG711, pictured in the Mach Loop during a dull afternoon on 27 October 2003. Being covered in masking tape it is not known what its status was at this time, but it clearly shows Operation TELIC markings complete with a Combat Air Wing badge on the tail, mission marks and nose art in the form of 'Oh Nell', in honour of Nell McAndrew, a model and media personality at that time. It was quite possibly awaiting a repaint hence the masking tape. This aircraft was written off almost exactly three years later on 24 October 2006, when the aircraft encountered a bird strike while on a bombing run on Holbeach Range, Lincolnshire. The crew ejected safely and were rescued while the aircraft crashed into mud flats in the Wash estuary. *(Scott Rathbone)*

GR.4 ZA585/AH, seen passing the north end of Thirlmere on 4 December during one of its first outings as the IX(B) Squadron 90th Anniversary jet. Of all the RAF Tornado GR 90th Anniversary schemes, this was probably the most disappointing.
(Scott Rathbone)

I've included this image purely to give an idea of the kind of weather that you could still see certain aircraft fly through, in fact I believe this was when I coined the phrase 'GR.4 weather!' At just after 11am on 11 December 2003, a three-ship of GR.4s made their way down the Thirlmere valley in the Lake District. The visibility was dire and at 400 ISO and f/4.5, my camera was only able to achieve a shutter speed of 1/60th second. This image is the only one of several that I took of the three aircraft to be remotely sharp and I have absolutely no idea of any of the serials of the three aircraft. *(Scott Rathbone)*

Above left: F.3(T) ZH552/ST of XI(F) Squadron based at RAF Leeming, passing Thirlmere on 6 February 2004. The Leeming based squadrons of XI(F) and XXV(F) were the most common in the Lake District part of LFA17. *(Scott Rathbone)*

Above right: Dismal days with early digital SLR cameras weren't much fun, but I always tried to make the most of it, even though I was mostly thwarted. One such time was the afternoon of 9 February 2004. Attempting to use anything above 400 ISO was practically impossible due to the noise produced, so it was generally a case of setting 400 ISO, a wide open aperture and hope to be provided with a fast enough shutter speed to at least get something. Here Tornado GR.4(T) ZD711/I of II(AC) Squadron based at RAF Marham, leads a pair of Marham based GR.4s through the Thirlmere valley, both aircraft being caught with a shutter speed of only 1/90th second, hence the quality not being great. *(Scott Rathbone)*

Left: The second of the pair of Marham based GR.4s on 9 February 2004 was GR.4(T) ZA410/DX in 31 Squadron markings. Pictured carrying an inert 1000lb bomb, this was also caught with a shutter speed of only 1/90th. *(Scott Rathbone)*

By 2004, unmarked, all-over grey GR.4s were becoming more common, as seen here with GR.4 ZA446, which was flying out of RAF Marham at the time. This aircraft met an earlier than expected demise after an engine fire caused it to divert to RAF Leuchars on 23 September 2009. Repair was not authorised and it was sent for storage at RAF Shawbury in 2010. It is seen here at Thirlmere on 27 February 2004 as one of a pair. *(Scott Rathbone)*

Joining ZA446 in the pair of GR.4s through Thirlmere on 27 February 2004 was GR.4(T) ZA541/III, wearing dual markings for II(AC) and 31 Squadrons. *(Scott Rathbone)*

Only seven minutes behind the pair of GR.4s in the previous two images on 27 February 2004 were a pair of 617 Squadron GR.4s, with ZA587/AJ-M, pictured here, flying as the number two of the pair. These were on a low-level training flight from their home base of RAF Lossiemouth. *(Scott Rathbone)*

From what I remember, 23 March 2004 was a relatively slow day in the Lake District, but you could usually count on something late in the day. Such was the case on this day, when with an hour or so of light left the scanner burst into life with 'Jackal flight' on the 12(B) Squadron air to air frequency. With a relatively strong audio it was obvious that they weren't too far away and sure enough, a few minutes later ZA367/FW came into view, a 12(B) Squadron GR.4(T). Knowing there were two from the scanner, we looked back for the second which was a mile or so in trail of the first. However, while concentrating on photographing the second, we were alerted to the noise of another aircraft which turned out to be a third in the flight, although this one was slightly higher and wider. Even so, it was nice to end the day on a high with a three-ship of RAF Lossiemouth-based Tornados. *(Scott Rathbone)*

ZA367/FW again as it heads north up Thirlmere, with a little touch of reheat. *(Scott Rathbone)*

The second aircraft of Jackal 1-3 on 23 March 2004 was GR.4 ZA543/FO, proudly displaying its 12(B) Squadron markings in some early spring colour. This aircraft went on to become the 12(B) Squadron 90th Anniversary special in 2005. *(Scott Rathbone)*

Seen passing through Dunmail Raise on 27 July 2004 is GR.4A ZG712/F of XIII Squadron. Although classified as GR.4A, the 'A' model reconnaissance equipment was actually now redundant and the aircraft flew as standard GR.4s with equipment isolated and left as ballast.
(Scott Rathbone)

Right: GR.4 ZA406/FU of 12(B) Squadron based at RAF Lossiemouth, pictured at Thirlmere on 2 August 2004. This was Mad Monday 2004 and although an above average day for traffic, it was a long way behind the numbers recorded in 2003. *(Scott Rathbone)*

Below left: GR.4 ZA597/AJ-O of 617 Squadron, seen with wings fully forward, something of a rarity at low-level. Seen in the Mach Loop on 9 August 2004. *(Brian Hodgson)*

Below right: II(AC) Squadron GR.4A ZA373/H seen passing through the Bwlch area of the Mach Loop on 12 August 2004. *(Scott Rathbone)*

GR.4 ZA613/AN, one of a pair of IX(B) Squadron Tornados through the M6 Pass on 17 August 2004. The M6 Pass fed into the Eden Valley and was often used as a more direct route to Spadeadam Range. *(Scott Rathbone)*

Right: GR.4(T) ZA595/VIII of XIII Squadron, seen in the M6 Pass on 17 August 2004. This aircraft suffered a bird strike in South East Scotland on 20 January 2014 and diverted to RAF Leuchars. It was declared a write-off and withdrawn from use. *(Scott Rathbone)*

Below left: Although not noticeable in the image, some dark conditions greeted ZD719/BS of 14 Squadron as it passed through the M6 Pass on 26 August 2004. *(Scott Rathbone)*

Below right: GR.4 ZA546/AG of IX(B) Squadron, fitted with an Air Launched Anti-Radiation Missile (ALARM), one of a pair from RAF Marham pictured at Dunmail Raise on 7 September 2004. *(Scott Rathbone)*

The second of the pair of Tornados from RAF Marham on 7 September 2004 was GR.4A, ZG713/G of II(AC) Squadron. This can also be seen carrying an ALARM.
(Scott Rathbone)

GR.4(T) ZG771/AZ of IX(B) Squadron, seen over Thirlmere reservoir on 15 September 2004.
(Scott Rathbone)

F.3(T) ZH553/RT was the lead of a pair of XI(F) Squadron F.3s through Thirlmere on 10 February 2005. I usually shot low-level jets at 1/500th second, but this is a classic example of where dull conditions early on had caused me to use aperture priority mode set to wide open, only for the conditions to change, with an aircraft turning up before I'd changed my settings again. *(Scott Rathbone)*

Next up we have a couple of images of the XIII Squadron 90th Anniversary special, carried by GR.4A, ZA401/XIII. This image shows it in full 67° swept-wing configuration in LFA7 on 8 March 2005. *(Paul Bunch)*

The second image of ZA401 sees it head-on towards the camera, again in LFA7 on 8 March 2005. *(Paul Bunch)*

Above: A rare event happened in LFA17 on 11 April 2005, when F.3 ZE839/XK was captured at low-level in full, live missile, Quick Reaction Alert (QRA) fit. It is thought that it was scrambled, but then subsequently cancelled en route and therefore decided on a low-level jaunt on the way home. *(Brian Hodgson)*

Left: Friday 22 April 2005 was an excellent day in the Lake District with twenty-two passes, this being exceptional for any day, let alone a Friday. Almost every aircraft in the RAF inventory capable of low-level flying made an appearance and this included a pair of XV(R) Squadron GR.4(T)s from RAF Lossiemouth. This image shows ZG754/TP passing in front of Helvellyn, one of the highest peaks in the Lake District. ZG754 was the aircraft involved in a fatal collision with a helicopter near Kendal on 23 June 1993, while flying from RAF Brüggen as a GR.1(T). The Tornado suffered substantial damage but was able to divert to Warton, sadly however, the helicopter crashed killing both occupants. *(Scott Rathbone)*

Early May 2005 saw XV(R) Squadron deploy three Tornados from their base at RAF Lossiemouth, to RAF Waddington for one week. The three aircraft all wore different markings and only included one wearing XV(R) Squadron markings, such was the way the aircraft were becoming pooled between squadrons. On 4 May 2005 a pair of the aircraft found their way to LFA17, taking in the route through Thirlmere, where we see the lead aircraft, GR.4 ZA462/AJ-P, which was the 617 Squadron 65th Anniversary aircraft. *(Scott Rathbone)*

The second aircraft of the pair of XV(R) Squadron GR.4s through Thirlmere on 4 May 2005 was ZD895/TI, the one aircraft of the three deployed to RAF Waddington marked for XV(R) Squadron. The third aircraft deployed to RAF Waddington carried 31 Squadron markings. *(Scott Rathbone)*

Above: Also passing through Thirlmere on 4 May 2005 was GR.4 ZD739, one of a pair from the Fast Jet and Weapons Operational Evaluation Unit (FJWOEU). The FJWOEU formed on 1 April 2004 with the merger of the Strike Attack OEU, F.3 OEU and Air Guided Weapons OEU. *(Scott Rathbone)*

Right: F.3(T) ZH553/DY of XI(F) Squadron, passing through Dunmail on 7 June 2005. XI(F) Squadron would disband at RAF Leeming in October 2005. *(Scott Rathbone)*

Earlier in the book we saw Tornado GR.1P ZA326 of the RAE based at Thurleigh airfield, Bedford. Some sixteen years later and now based at Boscombe Down with the Empire Test Pilots School (ETPS), ZA326 is seen at Thirlmere on 8 June 2005, during its final year of operation. *(Brian Hodgson)*

The earliest GR.1 airframe upgraded to GR.4 standard was GR.4(T), ZA365. It is seen here as AJ-Y with 617 Squadron, passing through Dunmail Raise on 10 June 2005. *(Brian Hodgson)*

Above: GR.4(T) ZG756/BX, the 14 Squadron 90th Anniversary aircraft, passing through Thirlmere on 25 June 2005. In my opinion the best of the Tornado 90th Anniversary schemes. *(Chris Chambers)*

Left: One of the small fleet of FJWOEU Tornado GR.4s was ZA609, seen here at Dunmail Raise on 7 July 2005. *(Scott Rathbone)*

Opposite: Another view of GR.1P ZA326 during its final year of operation, this time seen in the Mach Loop on 11 July 2005. It was generally seen more regularly in LFA7, along with other members of the ETPS. *(Kevin Wills)*

The Panavia Tornado at Low-Level • 113

F.3 ZG731 of the FJWOEU, seen in LFA7 on 11 July 2005. The OEU F.3s were generally quite rare at low-level. *(Paul Bunch)*

Mad Monday 2005 occurred on 1 August, only this year it was anything but memorable. From our location on Thirlmere, Tornados could be seen most of the morning routing east to west and vice versa, to the south of Grasmere, but it seemed obvious that they were giving Thirlmere a wide berth. At around midday we decided to make the move to the M6 Pass, inevitably missing a 617 Squadron GR.4, which made a pass through the valley as we made our way down from our location. The M6 Pass wasn't much better, with just a pair of 111(F) Squadron F.3s and this 617 Squadron GR.4, ZA556/AJ-C, although the Pass had seen some traffic prior to us arriving. A big disappointment after two consecutive years of decent traffic levels on this day. *(Scott Rathbone)*

Another of the RAF 90th Anniversary schemes was this one adorning GR.4 ZA564, the 31 Squadron anniversary aircraft, seen in the Mach Loop on 11 August 2005. Known as the 'Goldstars', the squadron insignia was displayed with pride and will feature later in the book with future special schemes. This wasn't the first time that ZA564 had carried an anniversary scheme, having had the honour of carrying the 27 Squadron 75th Anniversary scheme. *(Scott Rathbone)*

Also seen in the Mach Loop on 11 August 2005 was GR.4(T) ZA562, wearing dual XV(R) Squadron and 617 Squadron markings on the tail. By the time we left this location on this day there had been eight Tornado GR.4 passes, with eleven passes through the Tal-y-Llyn valley. *(Scott Rathbone)*

Over two years since the end of Operation TELIC and II(AC) Squadron GR.4A, ZA400, still sporting 'Scud Hunters' nose art, sharks mouth and combat wing badge from its time in the Iraq War, is seen in the Mach Loop on 11 August 2005. *(Scott Rathbone)*

The final image from the Mach Loop on 11 August 2005 is of GR.4 ZA601/O wearing II(AC) Squadron markings. This was its second pass of the day and was the only aircraft on this day to appear twice. *(Scott Rathbone)*

Some perfect late afternoon light greets GR.4A ZE116/X of XIII Squadron, as it passes through the Thirlmere valley on 17 August 2005.
(Scott Rathbone)

GR.4 ZA447/DE of 31 Squadron, seen at Thirlmere on 12 September 2005. This was one of three GR.4s from RAF Marham to have passed through Thirlmere on this day, reportedly flying low-level all the way from Marham to Tain range in the North of Scotland. As a GR.1 ZA447 was one of the more well known Operation Granby aircraft, coded EA with XV Squadron, it carried a sharks mouth and 'MiG Eater' nose art. *(Scott Rathbone)*

Having spent a quiet afternoon of 27 April 2006 at the entrance of the M6 Pass, we decided to make the move home. Having made the climb to the top (this location required a climb and descent to and from the photographic position) we heard a noise and looked back to see a pair of Tornado GR.4s heading into the valley. Still having my camera in hand I managed to grab a couple of shots, although not from a great position. What I did manage to capture of one of the pair shows ZG775/FB of 12(B) Squadron with an impressive inert load of ALARM and Brimstone missiles. The ALARMs are visible on the wing pylons either side of the external fuel tanks, while two of the Brimstone missiles are just visible under the left side air intake. *(Scott Rathbone)*

GR.4 ZA588/BB of 14 Squadron based at RAF Lossiemouth, passes through Thirlmere on 9 May 2006. *(Scott Rathbone)*

Another image from Thirlmere on 9 May 2006 sees GR.4(T) ZA604/TY wearing XV(R) Squadron markings. *(Scott Rathbone)*

Above: Closely following ZA604 in the previous image was GR.4A ZG714/Q, wearing XIII Squadron markings from RAF Marham. Being only a minute or so behind could mean it was in a flight with ZA604, or they just arrived at similar times, something that wasn't uncommon at the time. *(Scott Rathbone)*

Left: Completely unmarked Tornado GR.4s were becoming more and more common by 2006 and this is depicted in this image of ZG792, seen in LFA17 on 11 May 2006. This aircraft was written off on 27 January 2011 when an in-flight fire caused the crew to eject and the aircraft to crash into the sea off Loch Ewe, near Gairloch, Scotland. *(Brian Hodgson)*

GR.4 ZA449/W of II(AC) Squadron, seen passing through the Mach Loop on 8 June 2006. For many years the only variants seen wearing II(AC) Squadron markings were A and T models, but with the A model now defunct, standard strike GR.4s could be seen in both II(AC) and XIII squadron markings. *(Scott Rathbone)*

Above: F.3 ZE201/FB of XXV(F) Squadron based at RAF Leeming, looking resplendent in the sun as it passes through Thirlmere on 9 June 2006. *(Brian Hodgson)*

Left: GR.4 ZD747/AL of IX(B) Squadron based at RAF Marham, is seen at Thirlmere on 20 June 2006. This image clearly shows the larger 2,250lt external wing tanks often referred to as 'Hindenbergers', as opposed to the smaller 1500lt tanks. *(Scott Rathbone)*

Above: F.3 ZG797/GF of 43(F) Squadron 'Fighting Cocks' based at RAF Leuchars. This was the squadron flagship and was rarely seen at low-level, however it ventured down to LFA17 on 22 June 2006 and is seen here at Thirlmere. *(Kevin Clarke)*

Right: GR.4(T) ZA598/TW, seen wearing IX(B) Squadron markings but with an XV(R) Squadron code, who it was actually flying with at the time. It was seen in LFA20T during a low-level sortie from its home base of RAF Lossiemouth. *(Brian Hodgson)*

31 Squadron GR.4 ZA550/DD, pictured at Thirlmere on 11 July 2006. The aircraft is seen carrying a GBU-24 Paveway III laser guided bomb and Thermal Imaging Airborne Laser Designator (TIALD) pod. *(Scott Rathbone)*

When XI(F) Squadron disbanded at RAF Leeming in October 2005 some of their Tornado F.3s were split between the remaining squadrons, XXV(F) at RAF Leeming and 43(F) and 111(F) at RAF Leuchars. Some nine months after the end of XI(F) Squadron and ZE168/DN still carried its XI(F) Squadron markings and code despite now flying with XXV(F) Squadron. It was one of a four through Thirlmere on 11 July 2006, that appeared to be engaged in a 2v2 low-level intercept exercise, ZE168 being in the lead pair. *(Scott Rathbone)*

Following ZE168 and wingman, as part of the pair a minute or so in trail on 11 July 2006, was F.3 ZE342/FG. This aircraft is seen wearing a darker shade of grey than had normally been carried by the F.3, more reminiscent of the grey usually adorned by the GR.4, although they themselves were starting to appear in a lighter shade of grey. *(Scott Rathbone)*

GR.4 ZD716, flying in IX(B) Squadron colours but without a tail code, is seen passing through Thirlmere on 11 July 2006. *(Scott Rathbone)*

56(R) Squadron, the Tornado F.3 training squadron known as the 'Firebirds', were originally based at RAF Coningsby, prior to moving to RAF Leuchars when Coningsby was redeveloped to become the main Typhoon base. F.3 ZG793/WM is seen here at Thirlmere on 13 July 2006 carrying a more colourful tail than standard for the airshow season, 56(R) being the squadron that supplied the F.3 demonstration. *(Kevin Clarke)*

GR.4A ZG729/M of XIII Squadron based at RAF Marham, takes the corner towards Corris in the Mach Loop on 19 July 2006. *(Scott Rathbone)*

The next nine images are all from Thirlmere on 7 August 2006, otherwise known as Mad Monday 2006. This turned out to be the best Mad Monday since 2003, with fourteen GR.4 passes taking place between early afternoon and early evening. First up is a pair of GR.4s from 14 Squadron at RAF Lossiemouth, as they exit the valley over the dam at the north end of Thirlmere at approximately 13.15. *(Scott Rathbone)*

GR.4A ZA371/C of II(AC) Squadron, is pictured at 13.17, closely following the pair of GR.4s in the previous image. *(Scott Rathbone)*

At 14.54 GR.4 ZD746/TH made a pass through the valley. This aircraft was carrying an XV(R) tail code but while wearing markings for the long-since defunct SAOEU. It is likely that the aircraft had been in storage for some time prior to being brought back into service. *(Scott Rathbone)*

Around fifty minutes later at 15.46, the same pair of 14 Squadron GR.4s from 13.15 returned for another pass. As with the first pass, they both came through with wings swept and this image shows the number two in the flight, having swapped the lead from the first pass, as GR.4A ZG705/118, wearing XIII Squadron markings. This is the first image in the book displaying the new coding system, with each aircraft receiving a three-digit numbered code. Although this made it easier to identify individual aircraft, it was a sad time for squadron identities. *(Scott Rathbone)*

The second aircraft in the pair with ZG705 was GR.4(T) ZD741, unmarked and in the new lighter grey scheme. *(Scott Rathbone)*

GR.4 ZD709/FA was seen at 17.27, wearing 31 Squadron markings with a 12(B) Squadron code. *(Scott Rathbone)*

With the light now as good as it gets in the summer months, Tornado GR.4 ZD847/AA of IX(B) Squadron, is seen passing through Thirlmere at 17.39, one of a flurry of late passes. *(Scott Rathbone)*

Just five minutes later at 17.44, Tornado GR.4 ZD811 passed through, unmarked but with an unknown black 'R' on the tail.
(Scott Rathbone)

Left: The final image from a great day on 7 August 2006 sees GR.4(T) ZA551/AX in IX(B) Squadron markings, passing through at 18.12. With the aircraft mixed up between bases and pooled between squadrons, it was difficult to know which aircraft were flying with which squadrons or even from which base they had flown. However, it was nice to see Tornados still wearing full markings, as that would become more difficult to see as the years progressed. *(Scott Rathbone)*

Below: GR.4 ZG791/137 of XIII Squadron, passing through Thirlmere on 16 August 2006. *(Scott Rathbone)*

The 12(B) Squadron 90th Anniversary scheme was carried by GR.4 ZA543/FF, pictured here at Thirlmere on 16 August 2006. This scheme did take a little bit of criticism for the way the fox was depicted on the tail. *(Scott Rathbone)*

GR.4 ZD740/BG of 14 Squadron, pictured in the Selkirk to Moffat valley within LFA20T on 13 September 2006. *(Graham Farish)*

RAF Lossiemouth based GR.4 ZA453/022 in 12(B) Squadron markings, speeds through LFA17 during October 2006. *(Chris Chambers)*

GR.4A ZG709/V of XIII Squadron, pictured at 100ft in LFA20T on 14 November 2006. LFA20T is one of three Operational Low Flying (OLF) areas within the UKLFS, although their use is now somewhat limited. *(Tom Gibbons)*

A typical winter's day in the Lake District sees GR.4 ZA449/W of II(AC) Squadron, as it routes through LFA17 towards Keswick on 23 January 2007. *(Barry Swann)*

GR.4 ZA596/062 gleams in the low winter sun as it passes the north end of Thirlmere on 25 January 2007. This aircraft was written off at Kandahar, Afghanistan, on 20 July 2009, when a burner blow out during departure led to the crew abandoning the take-off. The speed and weight led to the aircraft overrunning the runway and the crew ejected suffering minor injuries, while the aircraft was engulfed in flames and destroyed. *(Scott Rathbone)*

F.3 ZE785 of 41(R) Squadron based at RAF Coningsby, seen in the Mach Loop on 25 January 2007. When 41 Squadron disbanded with Jaguars in 2006, the squadron number was passed on to the FJWOEU as a reserve squadron, becoming 41(R) Squadron. *(Richard Bland)*

Early morning sunlight could produce some interesting results at times if on the wrong side of the valley. GR.4 ZA550/042 in 31 Squadron markings, heads through Thirlmere on 6 February 2007 with its exhaust plume highlighted by the backlit sun. *(Scott Rathbone)*

F.3 ZE737/GK in 43(F) Squadron markings but thought to be on loan to BAE Systems at Warton, seen passing through Thirlmere on 14 February 2007. *(Scott Rathbone)*

An extremely dull day in the M6 Pass on 16 February 2007, sees unmarked GR.4 ZD890 speeding through during mid-afternoon. This was the only movement of the day and the lack of usable light necessitated the use of a slow shutter speed. *(Scott Rathbone)*

GR.4 ZD714/AJ-W wearing 617 Squadron markings but operating out of RAF Marham at the time, seen at Thirlmere on 26 February 2007. *(Scott Rathbone)*

F.3 ZG753/HH, the 111(F) Squadron flagship, one of a pair of special schemed F.3s through Thirlmere on 26 February 2007. *(Scott Rathbone)*

Joining ZG753 through Thirlmere on 26 February 2007 was ZG780, the F.3 marked up for the XXV(F) Squadron 90th Anniversary.
(Scott Rathbone)

F.3 ZE755/YL was a regular visitor to the UKLFS while operating out of Boscombe Down with QinetiQ. It continued to carry XXV(F) Squadron markings, but was used as a test and evaluation aircraft for the last few years of its life. It is seen here in the Mach Loop on 8 March 2007. *(Scott Rathbone)*

Also in the Mach Loop on 8 March 2007 was GR.4(T) ZA594 of II(AC) Squadron, one of a pair flying from RAF Marham. *(Scott Rathbone)*

F.3(T) ZH555/PT, with 111(F) Squadron flash on the tail, one of a pair of RAF Leuchars based F.3s passing through the Mach Loop on 20 April 2007. It is thought that they had been on an exercise in the south west and had made their way up through LFA7 on their way home. *(Scott Rathbone)*

Joining ZH555 in the pair of F.3s heading through the Mach Loop en route to RAF Leuchars, was another F.3(T), ZE786/TF, in 56(R) Squadron markings. *(Scott Rathbone)*

A pair of GR.4s storm towards Dunmail Raise on 1 May 2007. *(Scott Rathbone)*

At least once a month, a Tornado F.3 would venture down from its base at RAF Leuchars, enter LFA7, fly a circuit of the Mach Loop, then climb out and head home to Leuchars. This was the case on 22 May 2007, when F.3 ZE968/HB of 111(F) Squadron, flew through the Loop and then headed home at high level. *(Scott Rathbone)*

During late 2006 the Royal Saudi Air Force returned a number of their Tornado IDS aircraft to BAE Systems at Warton for upgrade under the Tornado Sustainment Program (TSP). Initially, two were upgraded to be used as development aircraft, one being ZK113 and this was thought to be the first time that one had been captured in the UKLFS since their return. It is seen passing through Dunmail Raise during the afternoon of 4 June 2007. *(Scott Rathbone)*

Having spent a full week up in the Lake District and LFA17 it had turned out to be a quiet week for Tornados. Adding to the TSP aircraft on the Monday were two pairs of GR.4s on Wednesday afternoon, with Friday morning producing passes by a pair of XXV(F) Squadron F.3s. One of the F.3s, ZE204/FC, is pictured here passing through Dunmail Raise on 8 June 2007. *(Scott Rathbone)*

The XIII Squadron 90th Anniversary GR.4A is seen again here, this time in the Mach Loop and in sunshine on 8 June 2007. ZA401 was probably the most commonly seen at low-level of all the 90th Anniversary schemes, although I personally only caught it once in backlit conditions. *(Brian Hodgson)*

GR.4(T) ZA549 is seen completely unmarked and in the lighter grey as it speeds through Thirlmere on 2 July 2007. *(Scott Rathbone)*

IX(B) Squadron provided GR.4 ZA469/029 to carry a special scheme to celebrate twenty-five years of the Tornado GR in RAF service and it is seen here in the Mach Loop with an unusual fit of ALARMs and no wing tanks. The date was 16 July 2007 and the aircraft was returning to RAF Marham via the Mach Loop having been on static display at the Royal International Air Tattoo (RIAT) at RAF Fairford. *(Scott Rathbone)*

Also passing through the Mach Loop on 16 July 2007 and also returning to RAF Marham from static display at RIAT, was another special scheme, this time GR.4 ZD748 in the II(AC) Squadron 95th Anniversary scheme, proudly displaying their SHINY TWO nickname. *(Scott Rathbone)*

A classic case of a lack of concentration came my way on 1 August 2007, when I forgot to change my camera settings having dropped the shutter speed for a C-130. When GR.4(T) ZA552/XI came charging around the corner, I was still set at a slow shutter speed, although fortunately I managed to get something reasonable of the XIII Squadron aircraft as it passed. *(Scott Rathbone)*

31 Squadron GR.4 ZA458/024, is seen passing through Thirlmere on 6 August 2007. *(Scott Rathbone)*

Colourful F.3(T) ZE735/TG, the 56(R) Squadron flagship, seen at Thirlmere on 9 August 2007. *(Graham Farish)*

GR.4 ZA591/058 of 31 Squadron, swings around the location known as Bluebell hill in the Mach Loop, on 4 September 2007. *(Scott Rathbone)*

II(AC) Squadron GR.4(T) ZA612/IV, also seen from Bluebell hill in the Mach Loop on 4 September 2007. *(Scott Rathbone)*

XIII Squadron GR.4, ZD792/100, seen in LFA17 on 10 September 2007. *(Graham Farish)*

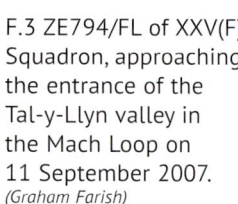

F.3 ZE794/FL of XXV(F) Squadron, approaching the entrance of the Tal-y-Llyn valley in the Mach Loop on 11 September 2007. *(Graham Farish)*

A pair of Tornado GR.4s route through Thirlmere on 26 September 2007. Both aircraft are in three tank fit, with an extra fuel tank under the fuselage, something that was commonplace during the 1980s and early to mid-1990s, but by this time had become rare to see. *(Graham Farish)*

A few minutes behind the GR.4 pair in the previous image was GR.4 ZA543/FF, the 12(B) Squadron 90th Anniversary jet. This was also in three tank fit and therefore easy to assume that it was in trail as part of a flight with the pair in front. *(Graham Farish)*

GR.4 ZA473, seen at Dunmail Raise on 1 October 2007. Wearing partial 14 Squadron markings, the aircraft also carries ALARM and Brimstone missiles. *(Barry Swann)*

The next four images come from two different locations within the Thirlmere valley on 4 October 2007. First up is GR.4 ZA393/008, one of a three-ship of Tornados based at RAF Lossiemouth, most likely flying with XV(R) Squadron. It is seen carrying an inert 1,000lb bomb. *(Graham Farish)*

Right: The weather during the morning of 4 October 2007 was atrocious, in fact those of us gathered at the north end of Thirlmere hadn't even got our cameras out, such was our belief that we wouldn't see anything. Fortunately for Graham he did have his camera out, and also managed to catch GR.4 ZA459/F, another of the three-ship of Lossiemouth Tornados. Also seen carrying an inert 1,000lb bomb, this was the current XV(R) Squadron 'McRoberts Reply', one of several to have carried this scheme and that are featured in this book. *(Graham Farish)*

Below: F.3 ZE254/FD of XXV(F) Squadron based at RAF Leeming, passing the north end of Thirlmere on 4 October 2007 in much improved weather. *(Scott Rathbone)*

Another view of ZE254/FD on 4 October 2007, as it turns topside to take the corner towards Bassenthwaite. *(Scott Rathbone)*

With the autumnal colour starting to take hold of the surrounding trees, F.3 ZE764/FK, one of a pair of XXV(F) Squadron jets, passes through Thirlmere on 16 October 2007. *(Scott Rathbone)*

Following ZE764 seen in the previous image on 16 October 2007 was the second aircraft of the pair, ZE961/FO. *(Scott Rathbone)*

Also seen on 16 October 2007 was GR.4 ZA611 of 41(R) Squadron. *(Scott Rathbone)*

XXV(F) Squadron F.3, ZE982/FR, is seen passing through the Selkirk to Moffat valley in LFA20T on 18 October 2007. *(Tom Gibbons)*

Another image of QinetiQ operated F.3 ZE755/YL, this time routing through Thirlmere on 6 November 2007. *(Scott Rathbone)*

With the Autumnal colours now in full swing, GR.4 ZA611 of 41(R) Squadron is seen passing through Thirlmere on 6 November 2007, surrounded by a wealth of colour. *(Scott Rathbone)*

F.3(T) ZE728/FZ in XXV(F) Squadron markings, passing through Thirlmere on a much duller 9 November 2007. The difference in the colours without sun is very noticeable. *(Scott Rathbone)*

F.3(T) ZE964/XT of 56(R) Squadron, makes a swept-wing pass through the Selkirk to Moffat valley on 13 February 2008. This valley has several photographic locations along its length. *(Tom Gibbons)*

Also making a swept-wing pass through the Selkirk to Moffat valley was F.3(T) ZH559/MT of 56(R) Squadron, this one being on 4 March 2008, one of my favourite Tornado F.3 low-level images. *(Tom Gibbons)*

As seen earlier, the RSAF TSP Tornados could be seen in the UKLFS during their time being upgraded at BAE Systems at Warton, usually the two development aircraft which both flew with Saudi markings blanked out and with the UK serials of ZH917 and ZK113. On 14 March 2008 one was captured in the M6 Pass in full RSAF markings complete with serial 6611. This was thought to be the second upgraded aircraft on a test flight, prior to being returned to Saudi Arabia. *(Tom Gibbons)*

With at least two RSAF TSP Tornados having returned to Saudi Arabia, the two development aircraft continued to fly regular test flights from Warton and on 18 March 2008, ZK113 was seen in the Mach Loop with a weapons fit of ALARM and Sidewinder missiles. *(Barry Swann)*

43(F) Squadron F.3 ZG757, passes through Thirlmere on 27 March 2008. This was still carrying a black spine and special tail for the 90th Anniversary of the 'Fighting Cocks', but when initially applied back in 2006, the aircraft was in a short-lived all-over gloss black scheme, something rarely seen carried by RAF frontline aircraft. *(Graham Farish)*

GR.4 ZD744/092 of XIII Squadron, makes its way around the Mach loop on 16 April 2008. *(Tom Gibbons)*

Unmarked GR.4(T) ZA602, sweeps into the Tal-y-Llyn valley in the Mach Loop on 13 May 2008. *(Brian Hodgson)*

GR.4 ZA469/029 in the Tornado GR 25th Anniversary scheme, makes its way through the M6 Pass on 22 May 2008. *(Scott Rathbone)*

43(F) Squadron F.3 pair of ZE887/GF and ZE764/GL, head towards Dunmail Raise from Grasmere on 13 June 2008.
(Scott Rathbone)

F.3 ZE887/GF, the new 43(F) Squadron flagship, seen routing through Dunmail Raise on 13 June 2008. *(Scott Rathbone)*

Another shot of ZE887/GF and ZE764/GL of 43(F) Squadron, at Dunmail Raise on 13 June 2008. As I've already mentioned, 43(F) Squadron was the only fast jet squadron that eluded me at low-level and the images from this day practically sum up my low-level collection of the squadron.
(Scott Rathbone)

Also passing through Dunmail Raise on 13 June 2008 was F.3 ZE168/FA. An earlier image of this jet in the book showed it flying with XXV(F) Squadron but in XI(F) Squadron markings after XI(F) Squadron had disbanded. This time it was pictured in XXV(F) Squadron markings but flying with 43(F) Squadron, after XXV(F) Squadron had disbanded on 4 April 2008. *(Scott Rathbone)*

GR.4 ZD848/TI, a IX(B) Squadron marked aircraft with XV(R) Squadron code, is seen over Grasmere within LFA17 on 20 June 2008. *(Graham Farish)*

BAE Systems development aircraft, GR.4A ZA402, passing through Thirlmere on 23 June 2008. It is seen carrying a converted ECM pod installed with cameras to capture weapons release trials. *(Graham Farish)*

In 2008 the by now renowned Mad Monday was 4 August and turned out to be a relatively good day with around ten GR.4 passes. Mixed conditions and the fact that most of the aircraft that day are already included in the book means that I've only included one from that day, that being ZD843/106 in XV(R) Squadron markings, another with a 56 Squadron Firebird sticker on the tail. It was seen at Dunmail Raise as one of a pair of XV(R) aircraft. *(Scott Rathbone)*

F.3 ZE757/FI of 56(R) Squadron, seen passing through Thirlmere while on a low-level training flight in LFA17 from its home base of RAF Leuchars on 8 September 2008. *(Brian Hodgson)*

A winter view of GR.4 ZA447 of 41(R) Squadron, as it heads to the Mach Loop from Bala through a misty valley on 10 December 2008.
(Graham Farish)

GR.4A ZA372 baring both its II(AC) Squadron code of 'E' and numbered code of '006', as it routes through LFA17 at Grasmere on 13 February 2009. *(Graham Farish)*

Of the three Tornado GR.4 development aircraft, two continued to wear camouflage from their GR.1 days, with coloured GR4 titles displayed. One, ZD708, was withdrawn from use in 2005, but ZG773 flew until 2016, although it did lose its camouflage scheme for the standard all-over grey scheme prior to its retirement. It is seen here passing through Thirlmere on 2 April 2009. *(Tom Gibbons)*

RSAF Tornado IDS ZK113 was back in LFA17 on 2 April 2009, seen as it runs through Thirlmere in some beautifully sunny weather. You can clearly see where the RSAF markings have been blanked out while flying test flights with BAE Systems. *(Tom Gibbons)*

Unmarked GR.4 ZA557/048, seen in the Mach Loop on 2 June 2009. *(Scott Rathbone)*

The next three images feature Tornado F.3s at Glen Coe on 10 June 2009. First we have ZE790/HC, one of a pair of 111(F) Squadron F.3s to pass through at 15.51. *(Graham Farish)*

Also making up the pair of 111(F) Squadron F.3s at 15.51 on 10 June 2009 was ZE200/HN, pictured here with air brakes partially deployed, adjacent to the bottom of the tail fin. *(Graham Farish)*

Giving an idea of how vast the valley is at Glen Coe are another pair of 111(F) Squadron F.3s, ZE288 and ZE734, just six minutes behind the previous pair at 15.57 on 10 June 2009. *(Graham Farish)*

An impressive image of GR.4 ZD707/077 of XIII Squadron, as it passes through the Bwlch area of the Mach Loop on 1 July 2009. *(Tom Gibbons)*

Anybody who has driven along the M6 Motorway through Tebay in Cumbria, will no doubt have seen the warning signs for low-flying aircraft. This image of GR.4A ZA395 on 6 July 2009, shows exactly why those signs exist. *(Graham Farish)*

A scenic view of GR.4 ZG779, as it routes north along Thirlmere on 23 July 2009. *(Graham Farish)*

The classic view from this location on Thirlmere of the waterfall and bridge, as GR.4A ZA369/003 of XIII Squadron passes through on a low-level sortie on 24 August 2009. *(Tom Gibbons)*

The 'Tremblers' of 111(F) Squadron were formed in 1917 and therefore, in 2007, F.3 ZE734/JU was given a special scheme to commemorate their 90th Anniversary. However, by the time of this image in LFA17 on 15 September 2009, the special scheme survived but without the titles carried to celebrate the anniversary, with the scheme seemingly retained as the squadron flagship. *(Brian Hodgson)*

GR.4 ZD844/107, seen passing through the M6 Pass on 15 September 2009. As a GR.1, this aircraft was involved in a collision with GR.1A ZA397 over Canada on 1 August 1994, while returning to the UK from a deployment. ZD844 was able to divert to the nearest airfield, but ZA397 crashed and was written off, with both crewmembers ejecting safely. *(Graham Farish)*

This is where the three-digit tail code came into its own, in being able to identify an otherwise completely unmarked GR.4, in this case GR.4A ZA370/004, looking good in the Autumnal late afternoon sun. *(Tom Gibbons)*

GR.4 ZD746/094 in 31 Squadron markings, pictured passing through the Mach Loop at Dinas Mawddwy on 14 October 2009. *(Graham Farish)*

Partially marked for 12(B) Squadron, GR.4 ZD715/083, passes Cammoch Hill to the north west of Pitlochry, during a low-level sortie in LFA14 on 2 February 2010. *(Tom Gibbons)*

Pictured in the M6 Pass on 17 February 2010 was F.3 ZE763/HD, one of a pair of 111(F) Squadron F.3s returning to their home base of RAF Leuchars, after a missile firing on Aberporth Range. *(Tom Gibbons)*

Following on from the previous image we see F.3 ZE983/HL, the second of the two 111(F) Squadron F.3s using the M6 Pass to RTB, after a missile firing on Aberporth Range on 17 February 2010. *(Tom Gibbons)*

F.3(T) ZE965/HZ of 111(F) Squadron, blends into a beautiful scenic image as we see it traverse through LFA20T above St Mary's Loch on 1 March 2010. *(Graham Farish)*

Another view of F.3(T) ZE965/HZ as it passes through a snowy Selkirk to Moffat valley above Loch of the Lowes on 1 March 2010.
(Graham Farish)

A pair of Tornado GR.4s head towards Dunmail Raise from Grasmere during a low-level flight through LFA17 on 12 April 2010. *(Graham Farish)*

GR.4 ZD793/101, unmarked but flying with XV(R) Squadron, seen at Glen Coe within LFA14 on 11 May 2010. *(Brian Hodgson)*

GR.4(T) ZD711/079, in what was by now the relatively common dual markings of 12(B) and XV(R) Squadrons, seen passing through Glen Coe in LFA14 on 12 May 2010. *(Graham Farish)*

Captured wearing II(AC) Squadron markings in LFA7 on 24 May 2010 is GR.4A ZG709/120. The eagle-eyed enthusiasts among you might recognise this as an A model by the Infra Red Line Scanner (IRLS) pod underneath the forward fuselage, but will have also noticed the missing Sideways Looking Infra Red (SLIR) window, presumably removed as part of a panel swap. *(Graham Farish)*

Unmarked GR.4(T) ZD812/104 pictured at the Bwlch exit in the Mach Loop on 3 June 2010. As mentioned earlier in the book with the image of ZD743/TQ on 4 August 2003, this was the other aircraft involved in the collision over the Moray Firth on 3 July 2012 while flying with XV(R) Squadron, with the loss of the crew of ZD743 and the pilot of ZD812, who died later in hospital after ejecting. *(Brian Hodgson)*

F.3 ZE201/HU of 111(F) Squadron, is seen from Helm Crag as it passes over Grasmere on the approach to Dunmail Raise on 5 July 2010. The aircraft was en route to its home of RAF Leuchars having departed RAF Waddington after attending the annual airshow. *(Graham Farish)*

GR.4 ZD749/097, seen here wearing IX(B) Squadron markings as it passes through the M6 Pass on 20 September 2010. *(Graham Farish)*

Heading in from the sea at Loch Broom within LFA14T is GR.4 ZA463/028, wearing 617 Squadron markings in beautiful surroundings on 7 October 2010. *(Graham Farish)*

ZD849/110, another 617 Squadron GR.4, not that it's as easy to tell with this one as it's dwarfed by the valley at Ballachulish, Scotland, having turned in from Fort William during a low-level training flight. It would proceed over Glencoe village (see next image) before entering the pass at Glen Coe on 11 October 2010. *(Graham Farish)*

Following on from the previous image and another view of GR.4 ZD849/110, as it makes the turn over Glencoe village within LFA14 to continue towards the Glen Coe pass on 11 October 2010. *(Graham Farish)*

GR.4A ZA395, pulling out of Glen Coe adjacent to Ben Nevis, the highest mountain in the UK, on 11 October 2010. *(Graham Farish)*

BAE Systems operated GR.4A ZA402, seen exiting the Bwlch area of the Mach Loop on 20 October 2010. As one of the two remaining development aircraft ZA402 wasn't allocated a three digit tail code. *(Tom Gibbons)*

31 Squadron GR.4 ZA452/021, passing through the Selkirk to Moffat valley at 100ft on 16 December 2010. This image gives a good indication of how well the light grey scheme can work to camouflage aircraft against the ground in snowy conditions. *(Graham Farish)*

LFA8 doesn't provide too many opportunities to capture low-level fast jets, but the Derwent Valley, famously used for training by the 'Dambusters' of 617 Squadron in 1943, is one such possibility. On 8 March 2011, GR.4A, ZA404/013 with 14 Squadron nose markings, is witnessed passing through the Derwent Valley during a low-level training flight. *(Tom Gibbons)*

Not long before the retirement of the Tornado F.3, ZE791 was adorned with an anniversary scheme to commemorate twenty-five years of the variant in RAF service. It is seen here passing through the M6 Pass in LFA17 on 28 March 2011, one of the final three RAF F.3s on their way to RAF Leeming for RTP, after the disbandment of 111(F) Squadron at RAF Leuchars six days earlier. *(Graham Farish)*

Also flying with ZE791 on 28 March 2011 was F.3(T) ZH554/HX, in 111(F) Squadron markings and complete with JU-C code to commemorate the 70th Anniversary of the Battle of Britain. JU codes were worn by the squadron during the Battle of Britain between 10 July and 31 October 1940. *(Graham Farish)*

Above: On 1 April 2010 41(R) Squadron merged with the Fast Jet Test Squadron (FJTS) previously based at Boscombe Down, to form 41(R) Test and Evaluation Squadron (TES). A pair of 41 TES Tornado GR.4s were caught in LFA8 on 29 June 2011, with ZA600/EB-G carrying a special scheme for the 95th Anniversary, leading the pair. *(Scott Rathbone)*

Left: The second of the pair of 41(R) TES Tornado GR.4s in LFA8 on 29 June 2011 was GR.4A ZG707/EB-Z. This former XIII Squadron aircraft carries two mission marks under the cockpit. *(Scott Rathbone)*

Jagdbombergeschwader 32 (JbG 32) was a Fighter-Bomber Wing of the German Air Force until its disbandment on 31 March 2013. The first squadron within the wing was 321 Squadron and having operated the Tornado IDS between 1984 and 1991, the squadron then re-equipped with the Tornado ECR in the SEAD role, becoming known as the 'Tigers' from 1996 until the squadron's disbandment on 26 October 2012. In 2011, Tornado ECR 4633 was given an elaborate special scheme to commemorate the 50th Anniversary of the NATO Tiger Association and, on 13 July 2011, it is seen taking a high line through the Mach Loop on its way to RAF Fairford to participate in RIAT 2011. *(Barry Swann)*

Dual 617 Squadron and 12(B) Squadron marked Tornado GR.4, ZD715/083, seen at Thirlmere on 12 January 2012. Note the large number of mission marks under the cockpit. *(Graham Farish)*

A case of the old and the new in this image of GR.4A ZA398, as this now redundant 'A' variant, originally converted for the reconnaissance role, is seen carrying a Reconnaissance Airborne Pod Tornado (RAPTOR), the latest reconnaissance technology that superseded the equipment carried internally by the 'A' model. While the GR.4A was now classed as a standard strike GR.4, RAPTOR could be carried by any GR.4, thus giving the whole fleet the same capabilities, with the only difference being the obvious visual modifications of the now obsolete equipment within the GR.4A. ZA398 is seen passing through Thirlmere on 28 March 2012 in the II(AC) Squadron 100th Anniversary scheme, which can also be seen in the next image. *(Neil Dunridge)*

The day after the previous image (29 March 2012) and Tornado GR.4A ZA398 is again seen passing through Thirlmere, although this time without the RAPTOR pod. This image gives a good view of the anniversary scheme and the 67° swept wing. This aircraft suffered a bird strike on 14 January 2014, which forced an emergency landing at Manston Airport, and was subsequently written off as a result of the damage. Ironically, it had also diverted to Manston after a bird strike just under a month earlier on 18 December 2013, although if its hopes were to retire there, it might not have liked its eventual fate of being allocated to the Defence Fire Training and Development Centre! *(Brian Hodgson)*

GR.4 ZA601/AJ-G of 617 Squadron, seen at Thirlmere on 4 May 2012. AJ-G was a significant code during the Dambusters raid of 'Operation Chastise', as it was the code carried by Lancaster B MkIII ED932, the aircraft flown by the leader, Wing Commander Guy Gibson. After a period of using single digit codes on Tornados during the 80s, the squadron returned to using the original AJ-* codes in the early 90s.
(Brian Hodgson)

Unmarked GR.4A ZG727/126, enters the Tal-y-Llyn valley in the Mach Loop during a low-level training flight through LFA7. *(Brian Hodgson)*

GR.4 ZD720/086, seen passing through Patterdale within LFA17 on 21 May 2012. *(Graham Farish)*

Despite a lot of unmarked GR.4s by this time, many did hang on to squadron markings, even though they often didn't reflect where they were flying from at the time. Fully marked for 31 Squadron, Tornado GR.4 ZD851/112, passes through the Mach Loop on 22 May 2012. *(Brian Hodgson)*

Tornado GR.4 ZD850/111 in 617 Squadron markings, seen passing through the A5 Pass in the north western part of LFA7 on 24 May 2012. Note the numerous mission marks below the cockpit. *(Barry Swann)*

The next two images show another RAF Tornado GR.4 special scheme aimed at the type rather than the squadron. In this case it was GR.4 ZA547 of XV(R) Squadron, with a scheme celebrating one million flying hours for the type. It was seen in LFA17 on 26 June 2012. *(Paul Massey)*

The other side of Tornado GR.4 ZA547 in the one million flying hours scheme, again seen in LFA17 on 26 June 2012.
(Gareth Jones)

The biannual, UK hosted, Joint Warrior (JW) exercises sees a number of air arms participate in what is essentially a maritime-based exercise usually centred in the north of England and Scotland, with aircraft deploying to a variety of bases that regularly included RAF Leeming, RAF Leuchars and RAF Lossiemouth. The exercise is held in April and October and October 2012 saw the German Air Force deploy fourteen Tornado ECRs of JbG 32 to RAF Leuchars for JW 12-2. The next four images display their use of the Scottish areas of the UKLFS during the deployment, with the first image showing 4646 taking a high line through Glen Tilt in LFA14 on 4 October, giving a good view of the AGM-88 HARM, the main weapon used in their role of Suppression of Enemy Air Defences (SEAD). *(Neil Dunridge)*

The next three images were all taken in the Selkirk to Moffat valley within LFA20T in the Scottish Borders on 8 October 2012. The first sees 4624 passing through and of note in this and the previous image, is the Cerberus ECM pod on the end of the left side wing and how different it is to the Sky Shadow pod used by the RAF. *(Neil Dunridge)*

Next we have 4646 diving into the valley on 8 October 2012. *(Neil Dunridge)*

The last of the four JbG 32 Tornado ECRs in the UKLFS sees 4649 heading west on 8 October 2012. This isn't the last we'll see of this aircraft in this book. *(Neil Dunridge)*

Showing signs of the withdrawal of the role of the 'A' variant and subsequent redeployment as a standard strike aircraft, GR.4A ZA404/013 is pictured in LFA17 on 28 February 2013 wearing 617 Squadron markings, one of the original strike squadrons that historically wouldn't have operated the 'A' variant. *(Brian Hodgson)*

The famous Dambuster raids over Germany by 617 Squadron during the Second World War, took place on 16 May 1943. For many years the anniversary was commemorated every five years, with flypasts by the BBMF Lancaster Bomber and 617 Squadron Tornados along Derwent Reservoir in the Peak District, the main area where the squadron trained for the raid. For the 70th Anniversary in 2013, two 617 Squadron GR.4s (ZA412 and ZA492) were given a special scheme and both participated in the flypast on 16 May 2013. This image shows ZA492 as it passes above the Derwent Dam during the anniversary flypasts. *(Scott Rathbone)*

GR.4 ZD790/099, passing through the Mach Loop on 30 May 2013. *(Barry Swann)*

41(R) TES GR.4 ZG777/EB-Q, powers through the Mach Loop during late May 2013. *(Scott Rathbone)*

617 Squadron 70th Anniversary GR.4 ZA492, seen passing through the Mach Loop in 67° swept-wing configuration on 3 June 2013. *(Barry Swann)*

A head-on view of 67° swept-wing GR.4, ZA550/042, as it enters the Tal-y-Llyn valley in the Mach Loop on 4 June 2013. *(Neil Jackson)*

Along with ZA492, the other 617 Squadron 70th Anniversary Tornado was GR.4(T) ZA412, seen here passing through Thirlmere on 18 June 2013. *(Brian Hodgson)*

GR.4A ZA401/012, seen in LFA17 on 19 June 2013 carrying a special scheme for the 30th Anniversary of the Tornado GR. *(Brian Hodgson)*

Above left: GR.4 ZD810/102, passing through Dunmail Raise on 6 August 2013. *(Andy Skarbinski)*

Above right: Unmarked GR.4 ZA463/028, entering the Tal-y-Llyn valley in the Mach Loop on 7 August 2013. *(Barry Swann)*

Left: Royal Saudi Air Force Tornado IDS 8306, seen at Dunmail on 5 September 2013. This was one of four RSAF Tornado IDSs that were deployed to RAF Coningsby alongside four RSAF EF2000s for Exercise Saudi-British Green Flag, which started on 4 September for two weeks. Unfortunately the RSAF had by now adopted an all-over grey scheme for their Tornados replacing the previous desert camouflage scheme. *(Andy Skarbinski)*

Above: With fully marked Tornado GR.4s becoming a rarity, it was nice to see ZA472/031 in full 31 Squadron markings through the Mach Loop on 22 November 2013. It is also seen carrying a Litening targeting pod. *(Scott Rathbone)*

Right: GR.4 ZD745/093 turns towards the Bwlch in the Mach Loop with the sun lighting up the exhaust in its wake on 22 November 2013. *(Scott Rathbone)*

Left: Another view of ZD745/093 as it passes through the Bwlch on 22 November 2013. *(Scott Rathbone)*

Above: Unmarked GR.4(T) ZD712/080 passing through Thirlmere 17 December 2013. *(Andy Skarbinski)*

Right: GR.4(T) ZD842/105 of XV(R) Squadron, caught in LFA17 on 30 January 2014. *(Andy Skarbinski)*

Below: Deep winter often leaves the valleys dark or in shadows by mid-afternoon, but late winter can produce some superb usable light at many locations within the UKLFS. Proving that on 26 February 2014 is GR.4 ZD747/095, seen passing through the northern part of the Mach Loop. *(Tom Dean)*

GR.4A ZA395 of 12(B) Squadron, carrying a special scheme for the impending retirement of the squadron. As mentioned in the introduction, the squadron was soon reinstated and would reach its 100th Anniversary the following year. ZA395 is seen here in the Mach Loop on 16 April 2014 on its final flight as it headed to RAF Leeming for retirement. *(Neil Dunridge)*

Also in the Mach Loop on 16 April 2014 was GR.4 ZD849/110 in 617 Squadron markings based at RAF Lossiemouth, although it was at the time operating with the RAF Marham squadrons. *(Scott Rathbone)*

Evening in the Mach Loop sees GR.4 pair ZA606 and ZD849 make their way through the Bwlch and into the left turn towards the Tal-y-Llyn valley. *(Barry Swann)*

Passing through the Mach Loop on 5 June 2014 was Tornado GR.4 ZA614/EB-Z of 41(R) TES. Another of what seemed like an abundance of GR.4 specials around the time, this one depicted Group Captain Donald Osborne Finlay, DFC, AFC, an RAF Officer and British athlete who competed at three Olympic games in 1932, 1936 and 1948, winning a bronze medal in 1932 and silver in 1936. The special scheme was in honour of his achievements and his command of 41 Squadron for a period between 1940–41 during the Second World War. *(Scott Rathbone)*

Above and opposite: Two images of GR.4 ZA606/069, both in LFA7 during June 2014. *(Scott Rathbone, above, and Tom Dean, opposite)*

41(R) TES GR.4, ZA614/EB-Z again, this time over St Mary's Loch at 100ft, as it performs an OLF flight in LFA20T on 17 July 2014. *(Andy Sheppard)*

GR.4(T) ZG752/129, passing gracefully through the Hawes valley in North Yorkshire as it routes through LFA17 towards Cumbria on 24 November 2014. *(Graham Farish)*

GR.4 ZA406/015, seen in LFA7 during February 2015. *(Brian Hodgson)*

Pictured at the entrance of the Selkirk to Moffat valley on the afternoon of the 23 April 2015 was XV(R) Squadron's centenary jet ZA461, operating as 'ALIEN 1', one of a pair making use of a timed slot within the TTA. XV(R) Squadron were at the time utilising the Scottish Borders on a regular basis, in particular the 100ft designated Low Flying Area LFA20T, as they pushed crews through their training prior to the squadron's disbandment. *(Andy Sheppard)*

GR.4A ZA370/004, pictured over Lake Windermere as it heads north through LFA17 on 14 May 2015. *(Graham Farish)*

GR.4(T) ZD742/090 of 617 Squadron, passing through Thirlmere on 10 June 2015. *(Neil Dunridge)*

Having seen anniversary specials for both twenty-five and thirty years, the latest RAF special adorned Tornado GR.4 ZD788, celebrating forty years since the first flight of the Tornado in 1974, is seen here at Thirlmere in LFA17 on 10 August 2015. *(Neil Dunridge)*

Seen the day after the previous image on 11 August 2015 and at the same location, was Tornado GR.4 ZA456, the IX(B) Squadron 100th Anniversary special. *(Neil Dunridge)*

Flying low over St Mary's Loch on 19 August 2015 is ZA589/057, undertaking an OLF sortie in LFA20T flown by XV(R) Squadron. *(Brian Hodgson)*

Another RAF special scheme, this time the 12(B) Squadron 100th Anniversary worn by GR.4A ZA405 and reminiscent of the squadron's retirement scheme in 2014, before their reprieve. ZA405 is pictured here on the eastern side of the Mach Loop on 13 October 2015. *(Graham Farish)*

The honour of carrying the fabulous 31 Squadron 100th Anniversary scheme fell to GR.4(T) ZA548, seen here in LFA20T in beautiful late afternoon light on 15 February 2016. Although a similar design to previous 31 Squadron specials, this one seemed to stand out more than the others, particularly in light like this. *(Tom Dean)*

In 2016 the RAF produced something that hadn't been seen for several years, a fast jet all-over special scheme. With many appearing during the 90s and even the early 2000s, generally all-over black and usually as air display schemes, it was something of a surprise to see GR.4(T) ZG750 rolled out for the 25th Anniversary of 'Operation Granby', in honour of their participation in the Gulf War and continued deployments, wearing an all-over desert pink ARTF scheme as carried during Desert Storm. It is seen here during its first known low-level flight around the Mach Loop on 18 July 2016. *(Scott Rathbone)*

A GR.4 with empty wing pylons just looks naked, but this is how GR.4(T) ZD713/081 appeared in the Mach Loop on 18 July 2016. It was seemingly on a test flight from the Tornado Combined Maintenance and Upgrade facility (CMU) at RAF Marham, using a 'Tarnish' call sign which is used by BAE Systems pilots. The CMU consisted of both RAF and BAE Systems personnel. *(Scott Rathbone)*

A rare event occurred on 10 October 2016 when GR.4 ZA554/046, was caught passing through the Bwlch area of the Mach Loop while a rainbow was present. Fortunately it hung around long enough for an aircraft to be captured against it, and although there are several images of this particular pass, I can't think of any other similar images taken in the UK. *(Brian Hodgson)*

Full head-on low-level images aren't as easy to capture as you might think. Even the best locations that provide an opportunity require a long lens and an element of cropping to produce reasonable results. The entrance to the Tal-y-Llyn valley in the Mach Loop is one such example, but even this image of GR.4(T) ZG750, approaching the entrance on 26 October 2016, was taken at 560mm and is still cropped significantly.
(Scott Rathbone)

The Operation Granby special schemed Tornado, ZG750, was affectionately known as 'Pinky' due to the desert pink colour scheme. To keep with tradition it carried a shark's mouth, something worn by several of the Tornado GR.1s during the conflict and also numerous mission marks, another symbolic reference to that adorned by the type during the war. It is seen here on 26 October 2016, a few seconds after the previous image, as it passes through the Tal-y-Llyn valley during a low-level sortie from RAF Marham. *(Scott Rathbone)*

Each year 41(R) TES deploy several aircraft to NAS China Lake, California, to take advantage of the good weather and vast weapons and test ranges. Nowadays it is just Typhoons, but when operational it included Tornado GR.4s. Not far to the north of China Lake lies Rainbow Canyon, part of the low-level complex known as the 'Sidewinder'. Unfortunately, due to an accident back in 2019, the canyon is no longer used, but back in its heyday, Tornado GR.4 ZA560/EB-Q, is seen passing through on 16 November 2016 during one such deployment to China Lake. *(Brian Hodgson)*

A relatively calm St Mary's Loch greets GR.4 ZA553/045 on 23 November 2016 as it routes through the Selkirk to Moffat valley during an operational low-flying sortie down to 100ft. *(Brian Hodgson)*

GR.4 ZA559/049 heads into the winter shadows of Thirlmere on 5 January 2017. *(Andy Skarbinski)*

Unmarked GR.4 (as with many airframes towards the latter end of service), ZD848/109, is pictured in East Lothian, part of LFA16, on 6 April 2017. *(Andy Sheppard)*

A view of the underside of a GR.4 as ZA462/027, charges into the Tal-y-Llyn valley on 26 April 2017 in full 67° swept wing. The three hardpoints are clearly visible under the belly, able to carry fuel tanks, weapons and pods. In this case only a Litening pod is being utilised. *(Scott Rathbone)*

The Tal-y-Llyn valley within the Mach Loop runs alongside the Cadair Idris mountain and enthusiasts refer to the entrance to this valley as Cad, short for Cadair. Both sides are accessible for photography, but the western side suffers with a lack of usable light until late in the day, despite providing the better angles. The next three images give an idea of the angles achievable from here. Firstly we have GR.4 ZG779/136 with fully swept 67° wing on 9 May 2017. *(Tom Dean)*

Next we have 'Pinky', GR.4(T) ZG750/128, passing through the tight entrance to the Tal-y-Llyn valley on 6 July 2017, again with fully swept 67° wing. *(Neil Dunridge)*

Left: The third image is of an unidentified GR.4, again with 67° swept wing and producing a nice vapour cloud as it pulls through the entrance of the valley in humid conditions on 10 July 2017. *(Dave Johnson)*

Below: For reference, this image shows how aircraft entering the valley look from the eastern side. Angles can vary on this side depending on where you position yourself, with around a quarter of a mile area to photograph from. The highest point tends to give underside views, but other areas can offer a variety of angles. 'Pinky' again features here as ZG750, now wearing its three-digit tailcode of 128, is seen passing through on 13 July 2017 en route to RIAT at RAF Fairford for static display, carrying three inert Paveway IV laser-guided bombs. *(Neil Dunridge)*

GR.4 ZA459/025, one of a pair of GR.4s to pass through the Mach Loop on 24 July 2017. *(Scott Rathbone)*

The final 'MacRoberts Reply' XV(R) Squadron special fell to GR.4(T) ZD741/F-LS, seen here in the Mach Loop on 29 November 2017.
(Tom Dean)

GR.4 ZA543 makes its way around the Mach Loop in some incredible light for photography, provided by sunshine and the reflective snow on 11 December 2017. *(Tom Dean)*

Photographed from the picturesque location of the Bwlch exit, GR.4 ZA553/045, makes the turn towards the Tal-y-Llyn valley to continue its circuit of the Mach Loop. It was one of a pair undertaking a low-level sortie in LFA7 on 18 December 2017. *(Scott Rathbone)*

An unusual backdrop greets GR.4 ZA607/EB-X of 41 TES, as it routes north through LFA17 on 16 May 2018. The town in the background is Windermere and the image was taken from the high ground to the west of Lake Windermere, a place often avoided by low-level photographers due to the width of the lake, however this image proves that it was worth trying. *(Gareth Jones)*

In 2016 GR.4(T) ZG771 received a special scheme for the 100th Anniversary of RAF Marham. Each side of the tail was slightly different and depicted all the aircraft operated from the base over the years. The left side is seen here as it passes through LFA20T on 7 June 2018, with the aircraft on this side commemorating the early years of the base to around the Second World War era. *(Brian Hodgson)*

Following on from the previous image, we see the right side of ZG771 showing aircraft from the Cold War era, as it passes through LFA17 on 26 June 2018. With less than a year of RAF Tornado operations remaining, RAF Marham would soon be starting a new era with a fifth-generation fighter in the form of the F-35B Lightning II. *(Neil Dunridge)*

On 1 April 2018, the RAF celebrated its 100th Anniversary. Numerous squadrons celebrated their own 100th Anniversary with special anniversary schemes, as did RAF Marham, as seen in the previous two images. It was hoped among enthusiasts that several more special schemes would appear directly for the main RAF anniversary, but alas, to the dismay of many, the best we got was a big red sticker on at least one of each type of serving aircraft. To say it was disappointing was an understatement, but the RAF aren't in the entertainment business, so I guess we should be thankful for what we do get. The honour (if you can call it that) of the Tornado GR.4s to carry a sticker fell to ZG752/129 and ZA554/046, which is seen here in LFA17 on 5 September 2018. *(Neil Dunridge)*

GR.4 ZD716/DH had the privilege of being the final 31 Squadron special when it was painted in a retirement scheme reminiscent of previous 31 Squadron anniversary schemes. It was photographed in the Thirlmere valley on 8 January 2019, which turned out to be the penultimate day of low-level flights by RAF Tornados in LFA17. *(Neil Dunridge)*

The final RAF Tornado to fly low-level in LFA17 fell to GR.4(T) ZA612/074, pictured at Thirlmere on 9 January 2017. For almost forty years LFA17 was a stronghold for the Tornado GR, as well as the F.2/3, but now the tell tale tone of an approaching GR.4 would be no more. However, as you will see, it wasn't quite the end of Tornado operations within LFA17. *(Brian Hodgson)*

The Aeronautica Militare (Italian Air Force) took part in Exercise Iniochos in Greece during early 2022, with three Tornado IDS and three Tornado ECR variants. Pictured on 29 March 2022 is an IDS variant, notable by the rectangular black markings by the wing roots, ECRs have double the length of the markings. Sand storms from the Sahara reduced the flying part of the exercise but some of the Tornado crews did manage to get familiarisation flying down to 500ft, the majority however didn't get below 1,500ft, their own countries' legal limit. *(Neil Dunridge)*

The UK hosts a large scale, biannual training exercise named 'Cobra Warrior', which has now been taking place for several years during March and September. Each exercise sees a number of NATO and international air arms take part for three weeks, generally deploying to RAF Waddington, RAF Coningsby and RAF Lakenheath. In September 2022, the German Air Force (GAF) deployed a number of Tornado ECRs belonging to Tactical Air Force Wing 51, to RAF Waddington and when not involved in exercise missions they utilised their time fulfilling other training needs, which included some low-level work. The morning of 16 September 2022 proved to be the best day to have caught them in LFA17, as they made use of some glorious sunshine at Ullswater, as seen here with 4649. *(Simon Pearson-Cougill)*

Another image of GAF Tornado ECR 4649, passing over Ullswater on 16 September 2022. *(Simon Pearson-Cougill)*

IDS MM7067/6-41 from 6° Stormo, 154° Gruppo at Ghedi, is seen here in the 67° swept configuration at 250ft AGL on 20 April 2023. The Italian crews made use of the good weather at Exercise Iniochos in 2023, practising their low-flying skills as low-level flying in Italy is restricted to 1,500ft AGL. *(Neil Dunridge)*

A classic look of a Tornado, 67° swept configuration! IDS MM7067/6-41 powers down a canyon in Greece during Exercise Iniochos on 20 April 2023. The Aeronautica Militare brought an increased mixed fleet of IDS and ECR variants to the exercise, building on a successful 2022. *(Neil Dunridge)*

Each year the Iniochos exercise is hosted by the Hellenic Air Force Tactics Centre at Andravida air base, Greece's fighter weapons' school. The visiting aircraft/crews also participate in their own continuation training using the low-fly system and ranges and IDS MM7014/6-13 from 6° Stormo, 154° Gruppo, is seen here in a valley heading back to base after visiting a nearby range on 20 April 2023. *(Neil Dunridge)*

The Aeronautica Militare deployed a number of Tornado aircraft from Ghedi Air Base, Italy, to 117 Combat Wing at Andravida Air Base, Greece, for Exercise Iniochos 2023. EA-200B (Tornado ECR) MM7051/6-72 of 6° Stormo 'Alfredo Fusco', 155° Gruppo, is pictured passing through the Gorge of Vouraikos at low-level during an afternoon sortie on 24 April 2023. *(Tom Gibbons)*

Deployed to 117 Combat Wing at Andravida Air Base, Greece for Exercise Iniochos 2023 alongside the Aeronautica Militare EA-200Bs (Tornado ECR) were a small number of A-200A (Tornado IDS) variants. With a rather grand Greek chalet as a backdrop, IDS MM7014/6-13 of 6º Stormo 'Alfredo Fusco', 154º Gruppo, lead aircraft of a pair, is pictured passing through the Gorge of Vouraikos during an afternoon sortie on 25 April 2023. *(Tom Gibbons)*

6° Stormo, 155° Gruppo, Aeronautica Militare Tornado ECR, MM7051/6-72, is seen at 500ft flying through the Greek countryside carrying an AGM-88 HARM missile, on 27 April 2023. The Tornado only has until 2025 with the AMI and the Squadron/Wing and SEAD role will be taken over by the Lockheed Martin F-35A. *(Neil Dunridge)*

RAF Tornado callsign list (selection of commonly used callsigns)

Callsign	Variant	Squadron/Unit
Abbot	GR.1/GR.4	TWCU/XV(R)
Adnam	GR.4	XV(R)
Alien	GR.1/GR.4	TTTE/XV(R)
Apex	GR.1/GR.4	TTTE
Apollo	F.3	OEU
Arcade	GR.1	II(AC)
Aston	GR.1/GR.4	TTTE/XV(R)
Axis	GR.1/GR.4	TTTE/XV(R)
Baldrick	GR.1	XIII
Banshee	GR.1	TTTE
Barrel	F.3	29
Batman	GR.1/GR.4	IX(B)
Bengal	GR.1	617
Beta	GR.1/GR.4	TWCU/XV(R)
Birch	F.3	V(AC)
Blackdog	GR.1	617
Blackhand	GR.1	17(F)
Blacknight	GR.1	17(F)
Blacksmith	F.3	43(F)
Blade	GR.1	TTTE
Bobcat	GR.1	17(F)
Brandy	GR.1/GR.4	TWCU/XV(R)

Callsign	Variant	Squadron/Unit
Brigand	GR.1	TTTE
Bulkhead	GR.4	IX(B)
Bullet	GR.1	TTTE
Buzzard	F.3	29
Carbon	F.3	V(AC)
Ceder	GR.1/GR.4	TWCU/XV(R)
Chappy	GR.1	TTTE
Charger	F.3	229 OCU
Chariot	F.3	XXV(F)
Cheiftan	GR.1/GR.4	II(AC)
Classic	GR.1	TTTE
Clayman	F.3	23
Club	GR.1/GR.4	TTTE/XV(R)
Clyde	GR.1	27
Cobra	GR.1/GR.4	XIII
Colt	GR.1	TTTE
Condor	GR.1	TTTE
Crusader	GR.1/GR.4	14
Crystal	GR.1/GR.4	TWCU/XV(R)
Cutlass	GR.1	TTTE
Dixie	GR.1	TTTE
Dulux	GR.1/GR.4	31
Eagle	F.3	XI(F)
Everest	F.3	RAF Leeming Wing/111(F)

Callsign	Variant	Squadron/Unit
Export	F.3	111(F)
Falcon	F.3	XXV(F)
Fang	GR.1/GR.4	IX(B)
Firebird	F.3	56(R)
Flint	GR.1/GR.4	TWCU/XV(R)
Focus	F.3	43(F)
Fresco	F.3	23
Gamma	F.3	XI(F)
Gamecock	F.3	43(F)
Gibson	GR.4	617
Horseman	F.3	229 OCU/56(R)/43(F)
Jackal	GR.1/GR.4	12(B)
Jacket	F.3	23
Javelin	F.3	XXV(F)
Jumbo	GR.1	27
Lancer	GR.1/GR.4	XIII
Laser	F.3	V(AC)
Lucky	F.3	229 OCU
Lynx	GR.1	TWCU
Macaw	GR.1/GR.4	TWCU/XV(R)
Maddog	F.3	229 OCU
Magnum	GR.1/GR.4	TWCU/XV(R)
Magpie	F.3	29
Mango	GR.1	TWCU

Callsign	Variant	Squadron/Unit
Mantis	GR.1/GR.4	XIII
Maple	F.3	V(AC)
Mentor	GR.1/GR.4	XV(R)
Merlin	F.3	XXV(F)
Midas	GR.1/GR.4	617
Minx	GR.1	TWCU
Mitre	GR.1/GR.4	TWCU/XV(R)
Monster	GR.4	31
Nighthawk	GR.1	17(F)
Nitro	F.3	XI(F)
Pewter	GR.1/GR.4	TWCU/XV(R)
Phoenix	F.3	56(R)
Poker	GR.4	XV(R)
Rambo	F.3	229 OCU
Ransack	F.3	29
Razor	F.3	23/XI(F)
Rocket	GR.4	31
Rooster	F.3	XI(F)
Saturn	F.3	229 OCU/56(R)/43(F)
Savage	F.3	XXV(F)/111(F)
Saxon	GR.1/GR.4	617
Scarab	GR.4	14
Scarlet	F.3	V(AC)
Scimitar	F.3	111(F)

Callsign	Variant	Squadron/Unit
Scorcher	F.3	229 OCU/111(F)
Scorpion	F.3	V(AC)/43(F)
Serpent	GR.1/GR.4	14
Snake	GR.1/GR.4	14
Spartan	GR.1/GR.4	II(AC)
Spike	GR.1/GR.4	TWCU/XV(R)
Staple	GR.1	TTTE
Stella	GR.1/GR.4	TTTE/XV(R)
Sultan	GR.1/GR.4	II(AC)
Tarot	F.3	XXV(F)
Tempest	GR.1	27
Tirpitz	GR.4	IX(B)
Torch	F.3	229 OCU
Totem	F.3	XI(F)
Trent	GR.1	27
Tribal	GR.1/GR.4	27/XIII
Triplex	F.3	29
Trojan	GR.1/GR.4	617
Vandal	GR.1/GR.4	617
Viper	GR.1/GR.4	14
Voodoo	GR.1/GR.4	31
Warlord	F.3	56(R)/43(F)
Wolf	GR.1/GR.4	12(B)
Zircon	GR.1	17(F)

RAF Tornado Base list

Base	Squadrons/Units based over time
Boscombe Down	TOEU
RAF Brüggen	IX(B), 14, 17(F), 31
RAF Coningsby	V(AC), 29, 229 OCU/65(R)/56(R), F3 OEU
RAF Cottesmore	TTTE
RAF Honington	TWCU/45(R)/XV(R), XI(B), XIII
RAF Laarbruch	II(AC), XV, 16, 20
RAF Leeming	XI(F), 23, XXV(F)
RAF Leuchars	43(F), 56(R), 111(F)
RAF Lossiemouth	12(B), 14, XV(R), 617
RAF Marham	II(AC), IX(B), 12(B), XIII, 27, 31, 617
RAF Mount Pleasant	1435 Flight

Acknowledgements

This book would not have been possible without the help of several people and each deserves my appreciation no matter how large or small their contribution. Firstly there's the photographers who have really made the book what it is, so many thanks to Adrian Walker, Andy Sheppard, Andy Skarbinski, Brian Hodgson, Barry Swann, Chris Chambers, Dave Johnson, Gareth Jones, Graham Farish, Ian Black, Kevin Clarke, Kevin Wills, Neil Dunridge, Neil Jackson, Paul Bunch, Paul Massey, Richard Bland, Simon Pearson-Cougill, Tom Dean, Tom Gibbons and Tony Paxton.

Thanks also to John Higgins for scanning almost all of Adrian's prints and for help with exercise and aircraft info alongside Barry Swann and Tom Gibbons. I'd also like to thank Tom for his help with proof reading and with some of the more technical RAF info.